In the Rainforest

IN THE RAINFOREST

Catherine Caufield

THE UNIVERSITY OF CHICAGO PRESS

The University of Chicago Press, Chicago 60637

 Published 1984
University of Chicago Press edition 1986
Printed in the United States of America

95 94 93 92 91 90 89 5 4

Published by arrangement with Alfred A. Knopf, Inc.

A portion of this work originally appeared in *The New Yorker.*

Acknowledgments for permission to reprint previously published material can be found on page 305.

Library of Congress Cataloging in Publication Data

Caufield, Catherine.
In the rainforest.

Bibliography: p.
Includes index.
1. Rain forest ecology. 2. Rain forests. I. Title.
QH541.5.R27C38 1986 574.5'2642 85-24620
ISBN 0-226-09786-2 (paper)

For my father

Contents

Acknowledgments

Innumerable people gave me information, advice, shelter, and encouragement during the research and writing of this book. I shall always be grateful to Jon Tinker for getting me started on tropical rainforests. Ira Rubinoff and Michael Robinson of the Smithsonian Institution, Tom Lovejoy and Michael Wright of the World Wildlife Fund, and Robert Goodland helped me tremendously in making contacts with scientists and development workers in the countries I visited. Without Cay Craig, Art Hanson, David and Ann Holdsworth, Jesus Idrobo, Cathy MacKenzie, Gloria Moreno, David Oren, Bruce White, Erna Witoelar, and many others who went out of their way to help me make sense of what I saw, my travels would have been useless—and a lot less fun. There are many others who helped me whom I do not list: Some I have mentioned in the book and it seems cruel to expose them twice. Others I cannot name for the sake of their jobs or their safety, and to them especially I send my gratitude and admiration. My thanks to Gill Coleridge and David Godwin for being encouraging, and to Chuck Elliott for being cautious. Lastly, because this book is about trees and the misuse of natural resources, I was glad to do my little bit to save paper by writing it on an Apple computer, and persuading (it wasn't difficult) my publishers to print it on 50 percent recycled paper.

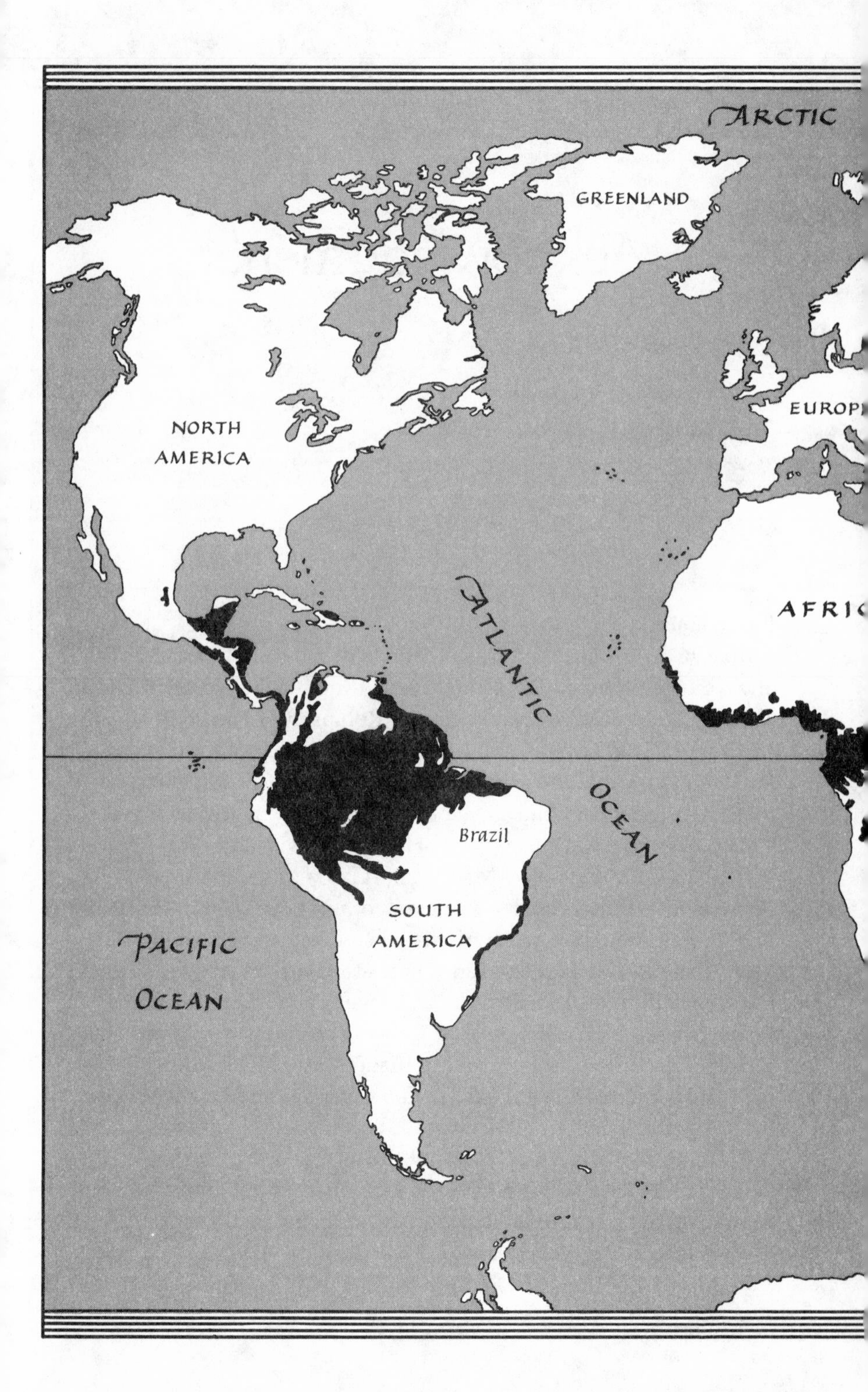
ARCTIC
GREENLAND
EUROP
NORTH
AMERICA
ATLANTIC
AFRI
OCEAN
Brazil
SOUTH
AMERICA
PACIFIC
OCEAN

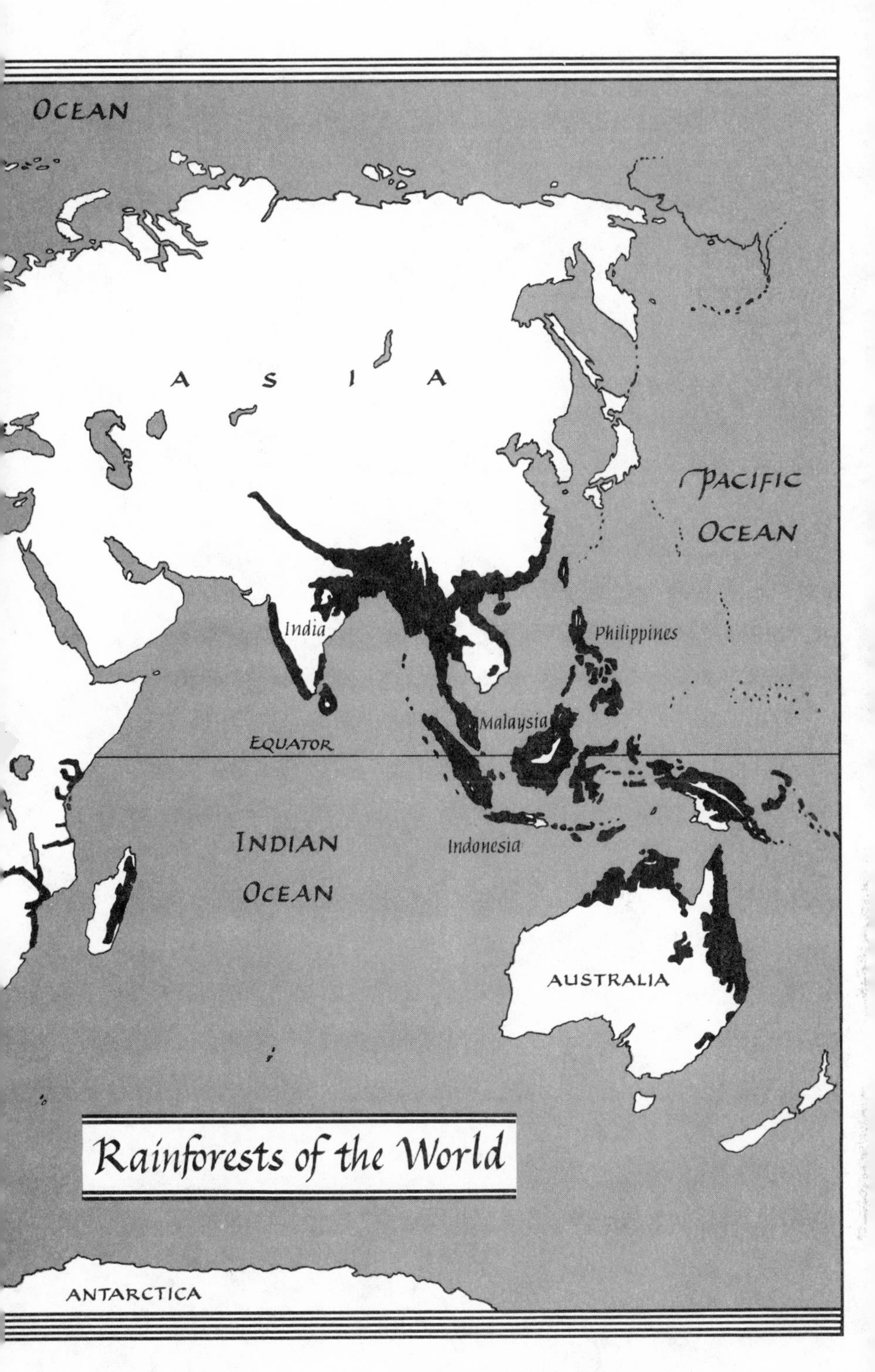
OCEAN
ASIA
PACIFIC
OCEAN
India
Philippines
Malaysia
EQUATOR
INDIAN
OCEAN
Indonesia
AUSTRALIA
Rainforests of the World
ANTARCTICA

In the Rainforest

1

A Town in Brazil

"I want to sell a city." His words, accompanied by an expansive gesture, hung awkwardly in the air. After a melodramatic pause, he continued, "I don't know if I can do it, but that's my dream. Brazil is a place where you can still dream." Tucurui, deep in the Amazon rainforest, is the city that Eduardo Albuquerque Barbosa dreams of selling. In 1977 there was no such place. Its projected site, on the banks of the Tocantins River, was home only to the plants and animals of the forest. Within three years air-conditioned offices, restaurants, bars, supermarkets, tennis courts, paved streets, houses, and fifty-two thousand people had changed all that. Eletronorte, northern Brazil's state electric company, built Tucurui to house workers constructing Brazil's largest dam. Tucurui Dam is the biggest engineering project ever undertaken in a tropical rainforest. When it is finished, in 1986, it will be the fourth largest dam in the world, only slightly smaller than the Aswan High Dam. The city of Tucurui, its purpose accomplished, will then be a ghost town, unless Barbosa's dream comes true.

Eletronorte has made an impressive job of housing, feeding, and otherwise caring for Tucurui's workers and their families. In many ways life here is much better than in an ordinary Brazilian town of comparable size. The company provides free schooling and medical treatment. There is a hospital with 220 beds, four operating rooms, and fifty doctors. The twenty-two schools have fifteen thousand students among them. Food is brought in by trucks and airplanes from São Paulo, three thousand miles away. Televisions, stereo equipment, washing machines, perfume by Yves Saint Laurent, makeup by Helena Rubinstein, and large quantities of bird cages, along with most of the rest of life's necessities, are available from the town's three supermarkets.

One of the most striking features of Tucurui is the sight of uniformed squads of laborers painting curbs, sweeping streets, and tending gardens. Everywhere one goes, and at all hours, groups of men or women are at work. My first sight upon arriving at half past nine at night was a gang of men in the street painting the curb white by the glare of the streetlamps. Another time I saw a mass of female streetsweepers, wearing matching brown trouser suits and hats, come down the road looking like something straight out of postrevolutionary Kampuchea. The result of all the primping and trimming is the creation of a genuine Eisenhower-era, suburban American dream. Building and maintaining this sterile paradise costs Eletronorte $3 million a month.

It is hard not to be impressed by the scale of things at Tucurui. The dam is 12 miles long. The reservoir behind it, when it fills, will flood 800 square miles of virgin rainforest. The countries of Grenada, Barbados, and Martinique could fit into the reservoir, with room to spare. Building costs have been running to $3 million a day. Workmen have already poured 8 million cubic yards of concrete into place, but before they could do that, Eletronorte had to build four refrigeration plants so that in the middle of the Amazon rainforest the concrete could be mixed at a maximum temperature of 53°F. Almost everything for the dam and for the people building it must be trucked in on the dusty, bumpy, rutted road from Brasília or floated down the Tocantins River from Belém, 180 miles to the north. For three or four months of the year, it is too wet to do much of anything. Nonetheless, Eletronorte has overcome most of the formidable logistical problems it faced in building the dam and its city.

According to Barbosa, Eletronorte, or at least the local management, is "trying to achieve a sort of *in camera* redistribution of wealth here." All the housing is owned by the company and assigned according to job category. Besides the huge barracks, in which the unskilled workers sleep, there are four standards of houses. The best, which rent for $20 a month, are built of brick and concrete and surrounded by high fences. At the bottom end of the scale are the temporary, prefabricated wooden houses, in which skilled laborers live. These, about one third the size of the "best" homes, cost $1.50 a month. Houses of a single type are grouped together in small clusters, which are in turn integrated with clusters of other types. In theory this means there is no segregation, of living quarters or education, since schools draw their students from the local neighborhood. It is, however, noticeable that the proportion of "best" houses rises as one moves toward the summit of the hill on which the new town of Tucurui is built, a nice indication of the social standing of their occupants.

Symbolically crowning the city, at the very top of the hill, stands Tucurui's best hotel, whose swimming pool, tennis courts, and air-conditioned bedrooms are reserved for visiting engineers, consultants, journalists, and other people Eletronorte wishes to treat well.

In the hotel's restaurant I had the first of many talks about the project with Barbosa, who has worked at Tucurui since 1979. He came as a statistician, but his proficiency in languages (he speaks English, French, and German, in addition to Portuguese) and his love of food, drink, and conversation made him a natural for the job of public relations man. He is a big man, over six feet tall, and built along the lines of Sidney Greenstreet. Like Greenstreet, he gives an impression of substance rather than fatness. By his manner and appearance, and his tendency to lecture me, I judged him to be in his late forties, though I learned that he was only twenty-nine.

Barbosa insisted that the town of Tucurui is an important social experiment. "We're trying to do things in a new way. We have to create a new nation—one based on the common man." I had trouble making political sense of his visions of a "new world." The notion of conducting experiments in social progress in a company town that will be almost entirely abandoned within three years struck me as almost a sinister delusion. Barbosa calls himself a socialist, but he despises Old World socialism. He sounded something like an idealistic army officer who believed that radical measures were needed to save his country from ruin. "Brazil should develop her own kind of socialism and lead the rest of the continent in a Latin American socialism based on our own cultures.

"You cannot push people. I have been asked by Europeans why there are no syndicates, labor unions, here. I tell them we can't set up unions for the workers. That's the way in the Netherlands maybe, but we believe it must come from the workers. They don't have syndicates because they don't see the value of syndicates. The first step is to have jobs, then syndicates."

The new city of Tucurui is six miles from a long-established riverside settlement, also called Tucurui. Before Eletronorte came to the area, about three thousand people, most of whom were usually absent hunting or fishing, lived in the old town. Now there are forty thousand people in old Tucurui, but still no sewers. Going from the new to the old town is like leaving a library and walking into a carnival. Suddenly there are lots of people on the streets, and they're not wearing uniforms. Every house is a shop, and every shop is wide open. The streets are lined with goods for sale—bicycle tires, mosquito nets, shoes, aluminum pots and pans, more shoes, clothes, fruits and vegetables, and shoes. The main street, and practically all the rest in what is now quite a big town, seems to be a bazaar.

Many of the houses have watermarks five feet up on their walls; quite a few are standing in what appears to be a lake. The water, I'm told, will recede in a month or two. Every year people get ready for the floods by taking everything out of their homes. After the water subsides, they move back, treating the annual exodus as an occasion for spring cleaning. The first impression is of cheerful chaos and great poverty. Considering that the vast majority of dwellings must have been built (if that is the correct way to describe the erection of shanties) in the last four or five years, they look incredibly ratty. The annual flooding evidently takes a severe toll.

Yet poor as it looks, old Tucurui is a boom town. It serves much the same function as did Bangkok for American soldiers on rest and recreation leave from Vietnam, and it has something of Bangkok's air. Catering to the traditional needs of the Eletronorte workers has been good for at least some of the residents. Through the open doors of a one-room shack comes the sound of a television, accompanied by a glimpse of five children sitting on a bed and staring straight ahead with the glassy-eyed expression common to television addicts everywhere. Stereo music blasts forth from sleazy lunch bars that look like they haven't seen a customer for years.

About half the current population of old Tucurui moved there because their homes were in the area that was to be flooded for the reservoir. Eletronorte gave the families that had to move a choice of going to the town or relocating in the countryside. Sixty-five percent chose to leave rural life behind. Many of those who wanted to stay on the land have since followed the others to town, their property expropriated by powerful ranchers using bribes, threats, and violence.

The Tucurui region exhibits a pattern familiar throughout the Amazon whenever new areas are settled. At first many peasants work their own small plots, but within a few years a small number of powerful landlords control most of the land. One factor in this pattern is the utterly chaotic state of property titles in the Amazon, with conflicting, inaccurate, and unregistered claims making it almost impossible to establish beyond a doubt who owns what. Well-connected businessmen naturally have a better chance of holding onto prime properties than illiterate farmers, regardless of the merits of a case. But the wealthy have a still more direct way of getting what they want. It remains common in the Amazon for businessmen to employ *pistoleiros* (hired gunmen) to get land for them. A few hundred dollars puts ten or twelve armed men at your disposal.

The contrast between old and new Tucurui is a boon to critics of Eletronorte and of the government's Amazon policy. Colorful the old town may be—and the new town may be sterile—but the squalor and poverty of the former is in shocking contrast to the tidiness and luxury of the latter. "We have a big problem," Barbosa told me. "Here is the new town all neat and clean, and the old town looks like shit. We get a lot of criticism."

This, so far as Barbosa is concerned, is all cruelly unfair. Eletronorte had wanted to build housing for its workers in and around the original riverside settlement. That way the company reckoned to avoid just the kind of criticism it is now getting—and lower its costs. Its plan was for local people to set up the shops, restaurants, and other enterprises Eletronorte would need. The company would have helped get them started and installed public facilities, such as sanitation, water, and roads.

"We wasted a year trying to get permission for this from the government agency in charge of development in the Amazon and the state planning department. We even built one housing estate right by the old town. We figured that would make the authorities see the thing as a fait accompli. But we couldn't get the go-ahead, and we were losing workers because there was no place to live. There were no schools, no facilities. So we went ahead and built elsewhere." The social and economic gap between the old and new towns is, according to Barbosa, "the biggest tragedy of the project." The ten-foot-high wire fences and the guarded gates of the new town do not bring the two closer together.

Eletronorte facilities are for residents of the new town, though in some circumstances the company extends the privilege (never the right) of using them to others. The hospital takes emergency cases from the old town. Any child from old Tucurui can go to high school in the new town (there are no high schools in the old town), but this is an offer that is seldom taken up. Few families in the old town can spare teen-age children to go to school, even if their lower-school education made it academically possible. The shops and bars "are open to anyone who can afford them," and who can make it past the well-guarded entrances to Eletronorte's model city.

Eletronorte provides some services to the old town. Officials supply food to the hospital there. Every year they help combat the aftereffects of flooding in the town. And they have paved the roads in the old town. "In fact we've paved them four times." To look at them, one would never suspect that the dusty, deeply rutted streets in old Tucurui had been asphalted even once, let alone four times. The problem is that small cracks in the asphalt have to be filled in as soon as they appear or they grow—helped by the intense heat and heavy rainfall—until the surface simply caves in. In the new town, with its gangs of uniformed workers, the roads are properly maintained. But, with no middle class, no administrators, no engineers, no tax base, and virtually no federal or state support, old Tucurui cannot even take proper care of the handouts it gets from Eletronorte.

In any case, those handouts will stop abruptly when Eletronorte dismantles its prefabricated houses, pays off Tucurui's teachers and doctors, and leaves only a skeleton staff behind to run the dam. Like a parasitic plant whose host has been uprooted, old Tucurui is bound to suffer. There is a chance that some industrial group that needs an Amazonian headquarters will take over the new town. That, as he said, is Barbosa's dream. "But even if the city finds a buyer, I'll go. I want to work on the Santa Isabel Dam next. That's going to be a great one." His eyes lit up at the prospect of another Amazonian experiment. "Dam people don't like to stay in one place." Three years ago Barbosa wasn't a dam person. He was a city planner with a bachelor's degree from the University of Brasília who had worked in Rio de Janeiro. "Oh, but I became a dam person in the first six months here," he says, laughing.

. . .

At the foot of the hill from which rises the new Tucurui, seventeen thousand laborers live in barracks. Apart from a few women who work as streetsweepers or in the kitchens, they are all men and, at least for the duration of their stay in Tucurui, all single. These are the unskilled workers building the dam. Most of them are from the poverty-stricken northeast of Brazil. For many this is their first job. Those from rural areas would have hunted and, if their families were lucky enough to have some land, farmed. The men and boys (several hundred are under eighteen years old) who come from Recife and others of Brazil's desperately poor urban slums are likely to have more experience of scavenging or begging than of paid employment.

The pay for an unskilled laborer at Tucurui is 22,000 cruzeiros (about $110) a month, 70 percent above the legal minimum wage in Brazil. Food and lodging is subsidized. But even with good wages here and the lack of opportunities at home, turnover among unskilled workers is high. Every year about 70 percent leave. At the beginning, when there were fewer amenities, turnover was even worse. Now there is further encouragement to stay in the form of a sliding scale of subsidies. For the first three months a worker pays 2,000 cruzeiros a month for food. Subsequently, the price goes down every three months, until after one year meals are free. The men's leisure time is devoted to sleeping, drinking beer (nothing stronger is allowed through the gates), and playing soccer. Less wholesome fun is available in the old town.

The boredom and loneliness of barracks life occasionally erupts spectacularly. Two years ago something happened in Tucurui that was straight out of a Hollywood movie. It was Easter Week, and a holiday. Some of the workers decided to observe the Holy Saturday tradition of parading an effigy of Judas Iscariot through the streets. The company watchmen decided—not without reason, by all accounts—that the figure was obscene. They interrupted the procession; there was a scuffle, and six men were taken to the town jail.

The rest of the men went to the canteen, the company mess hall, and spread the story. Suddenly the scene was Jimmy Cagney in the prison dining room in *White Heat,* learning of his mother's death. The men passed messages down long trestle tables. They banged their spoons against the metal plates. Food flew across the room. Suddenly 1,500 men began ripping the canteen apart. They overturned tables and tossed chairs through windows. Next they stormed the jail and got their colleagues out. After that, the mob swarmed into one of the supermarkets and began sweeping things off the shelves. Bags of flour, rice, and beans; cans of oil; and neatly stacked cartons of yogurt spilled everywhere. Several of the shelves held bottles of alcohol—Johnnie Walker Red, Bombay gin, Moët et Chandon champagne, Southern Comfort. These did not go to waste.

The terrified Eletronorte watchmen, who do not carry guns, called in the federal police, who do. The police fired their machine guns into the air—and into one of the rioters. The melee subsided. Eletronorte officials, terrified of the scandal that would erupt if one of the workers died as a result of police machine-gun fire, chartered a jet to fly the injured man and two of his friends to a hospital in São Paulo. "We didn't want any questions of his not getting proper treatment," one official explained. "We sent his friends with him as witnesses that he was getting the best care possible." Luckily the man recovered.

. . .

In April 1982, during Brazil's National Indian Week, Eletronorte arranged an exhibition in the local gymnasium of the cultural artifacts of the people whose ancestral lands would be flooded by the dam. Agents from the government's National Indian Foundation (FUNAI) presided forlornly over card tables on which artifacts of the Parakanan culture were neatly laid out with price tags. Many of these carried a message about the Parakanan's view of the world, but the symbolism was obscure to the local schoolchildren who were marched through the show. They gravitated to what they already knew of the Indians. Bows and arrows and spears, many of them beautifully decorated with feathers, drew crowds of boys. The girls held necklaces up to themselves for their friends' approval. Some of the feathered ornaments were stunning. One headdress made of black-and-white-striped hawk feathers forming a semicircle four feet across was on sale for $15. Barbosa showed me around the exhibition. "The anthropologists are shocked and say we are commercializing the Indians by selling their crafts. They have an ideal of an untouched Indian culture. But the Indians need money. They need food. They are making these to sell. To preserve Indian culture, the first step is to preserve the Indians."

Six towns had to be abandoned and six thousand families were forced to leave the area that Eletronorte decided to flood. Three reserves that had previously been set aside for the Parakanan and the Gavioes Indians will be affected; one will be almost completely flooded. The second reserve will be partly flooded, with a new section of the Trans-Amazon Highway passing through the rest of it. The third, which in the late 1960s was divided in two by the construction of a state highway, is in the pathway of an electrified railway and a high-tension electrical transmission line.

The Brazilian authorities tried many times during the first half of this century to "pacify" the Parakanan, who then numbered about one thousand. The Indians hindered the construction through their lands of the Tocantins Railroad, which was built to take brazilnuts safely to market by bypassing the river's treacherous rapids. (The railroad, which took more than forty years to complete and cost many lives, was abandoned within twenty years when the government put a road through nearby.) Pacification is a technique for getting the confidence of tribal people, most of whom are reluctant to show themselves to outsiders. The person who wishes to make contact leaves small trinkets—mirrors, fishhooks, beads—at a clearing in the forest and withdraws to give the group time to examine them. If all goes well, they accept the "gifts." The pacifier puts out more things and waits for the group to take them. This process goes on for as long as necessary, until the "pacified" people finally come out to meet their benefactors.

For fifty years the Parakanan resisted all such advances, but in 1953 the government's Indian Protection Service (SPI) managed to convince 190 of them to settle near its newly established Indian post in the area. SPI agents told the Indians they were bringing protection. In fact, they brought death. Influenza is a trival illness to whites, but for Indians it is one to which they have no natural resistance. Within a year of their pacification over 50 Parakanan had contracted flu and died. The rest, many of them sick and dying, escaped into the forest, out of reach of the SPI.

For the next fifteen years the Parakanan hid from outsiders. During that time the SPI was replaced by a new agency, FUNAI. The SPI was disbanded in 1967, after a government investigation revealed that Indian agents had practiced genocide by deliberately introducing smallpox, influenza, tuberculosis, and measles into Indian groups and had joined with speculators and landowners in systematic murder and robbery of their charges. Brazil's attorney general, Jader Figueiredo, who carried out the investigation, said that "by the admission of sexual perversions, murder and all other crimes listed in the penal codes against Indians and their property, one can see that the Indian Protection Service was for years a den of corruption and indiscriminate killings."

In early 1971, in preparation for the construction of the Trans-Amazon Highway, destined to pass through the Parakanan's land, FUNAI successfully pacified another group of Parakanan. It took eight months to overcome their wariness. Within weeks of their pacification forty of the group contracted influenza. Later the same year a Brazilian doctor, Antonio Madeiros, visited the Parakanan. He discovered that thirty-five of the women had venereal disease, an ailment hitherto unknown to the Parakanan. Madeiros found strong evidence that they had been infected by two local FUNAI agents and various highway workers. Eight children later born to those women were born blind. Eventually two of the women also went blind, as a result of their illness having gone untreated.

In July 1971 FUNAI established a reserve for the Parakanan on the border of the Trans-Amazon Highway. Nonetheless, the Aborigines Protection Society, which in 1972 sent a team out from London to investigate the conditions of the tribes of the Amazon Basin, found that highway workers were freely entering the Parakanan reservation, that the workers were not given medical examinations to ensure that they did not pass on diseases to the Indians, and that they were using intimidation and bribery to impose their will on the group. The group had had 150 members at the time of contact in 1971. By the end of the year almost half were dead, victims of influenza, malaria, and psychological disruption.

The team from the Aborigines Protection Society found the Parakanan in a state of cultural disintegration. They had been moved from their homeland and could not support themselves in the new area, partly because the disruption caused by the road building inevitably drove the game animals away. They were becoming dependent, as a consequence, upon handouts and bribes from FUNAI agents and highway workers. Many of the tribe had given up their traditional dress, which was well designed to withstand the rough life of the forest, for ill-fitting hand-me-down cotton rags that fell apart after a few months of outdoor life. A large proportion of the community was constantly under par, with colds, runny noses, and a general condition of weakness due to malnutrition. Highway workers introduced drink and prostitution, along with various infectious diseases that from time to time reached epidemic proportions. In 1975 Eletronorte announced that the Tucurui Dam would involve flooding a large part of the Parakanan reserve and rerouting a section of the Trans-Amazon Highway through what was left of it.

In 1976 FUNAI pacified another group of about 38 Parakanan who had had no previous contact with outsiders. After pacification FUNAI vaccinated everyone in the group and airlifted them 200 miles to a reserve called Pucurui, although most of it was already scheduled to be flooded for the new dam. In the mid-1960s the reserve had been bisected from north to south by the Belém–Brasília Highway. In the early 1970s the Trans-Amazon Highway divided it from east to west, and a busy town grew up at the junction of the two roads, inside the reserve. This group went through the same miseries that their cousins had undergone five years earlier. Transferred from land on which they had been able to support themselves to a "reserve" that closely resembled a traffic island, they suffered from physical and spiritual malnutrition. Within a year of being "pacified," ten members of the group had died of disease. The nearest medical assistance is the Eletronorte clinic at Tucurui, 80 miles by road from the Parakanan reservation and more than 40 from Pucurui.

The area now known as the Pucurui Reserve must once have been rich in game. But it is now largely barren—overhunted by workers from two nearby farms and the large FUNAI sawmill complex in the reserve, who use high-powered shotguns and modern traps. The noise of the sawmill and the consequent forest destruction also make the reserve less attractive to game. Collecting and selling brazilnuts could provide the Parakanan with a decent income, enough to reduce their dependence on handouts from FUNAI and enable them to replace their disappearing traditional sources of protein, but instead the valuable nuts are gathered and sold by FUNAI employees. The sawmill is processing wood from the soon-to-be-flooded forest. In 1976 the value of this wood was estimated to be over $50 million. By law the products of Indian lands belong to the Indians, but FUNAI has made no arrangements to return the sawmill profits to the Parakanan.

Both groups of Parakanan reacted to the disruption of their lives in a way which is classic in Indian cultures. They chose to commit collective suicide. They stopped hunting, planting crops, and having babies; they simply waited to die.

. . .

"In the bad old days, which were not so long ago," one observer recently remarked, "an army loudspeaker truck would go through the affected villages a couple of weeks before the reservoir waters reached them and say, 'Listen fellas, you've got two weeks to hop it. Goodbye.' " What happened at Tucurui does not seem much changed from those bad old days. The idea of a dam there had been discussed since the early 1960s, and it was well known that there were Indians in the region of the planned dam. Nevertheless, neither FUNAI nor Eletronorte made any provisions for indemnifying or relocating them until 1977, when construction was already underway. By then both groups of Parakanan were almost too demoralized to care what happened to them and afraid to count on FUNAI's promises of a secure future. But in 1978, facing the need to find a new place to live, each group chose to return to a different part of their ancestral lands. As soon as they were assured they could do so, three Parakanan women became pregnant. Both resettlement areas, however, are plagued by squatters and neither has been officially set aside by the government as an Indian reserve.

The Gavioes live in Mãe Maria, the third reserve to be affected by Tucurui Dam. They were "discovered" by explorers early this century, but for fifty years they successfully avoided non-Indians. Following their eventual pacification in the 1950s, 70 percent of the Gavioes died. By 1961 there were fewer than thirty Gavioes left in the world. A Brazilian anthropologist, Roberto Da Matta, who visited them, noted in the mid-1960s that the Gavioes at that time began to disappear as a group. "From then on they were to be a mere collection of individuals totally dependent on the national society."

But by 1977 the Gavioes had made an astonishing comeback. Their population was 108. They had driven the FUNAI agent off their land and had assumed control of their own affairs. In the intervening period the Gavioes had been moved a few at a time from their traditional lands to the Mãe Maria Reserve, which in the late 1960s was cut in two by the building of a state highway. In 1971, when a deputation from the Primitive Peoples' Fund and Survival International visited one of the Gavioes' villages, "every single woman and child was suffering from what appeared to be whooping cough, and all were sitting dejectedly with running noses, coughing ceaselessly. When asked what they were suffering from, the FUNAI official said, 'Nothing. It is the normal state for Indians to be in.' "

Nevertheless, Mãe Maria had one great advantage for the Gavioes. One of the Amazon's most valuable species, the brazilnut tree, grows there in abundance. In 1972 brazilnut exports brought the Tucurui region alone a profit of over $13.5 million. Since 1964 the main activity of the Gavioes has been the gathering of brazilnuts. But for most of that time the Indians were obliged to sell the nuts to FUNAI, which marketed them and gave the Gavioes 20 percent of the price, sometimes less.

In 1975 an anthropology student, Iara Ferraz, was brought in by FUNAI to run a "community development project." She encouraged the Gavioes to take control of the transportation and marketing of the nuts, as well as their collection. In June 1976 they dismissed the

FUNAI agent from the reserve and refused to sell any more nuts to the agency. Instead, they sold directly to the wholesalers and used the profits to buy a truck. FUNAI dismissed Ferraz from her post for alleged indiscipline, but the following year the Gavioes negotiated a loan with the Banco do Brasil to expand their operations, and hired twenty Brazilians as day laborers. By 1980 they were earning over $50,000 a year from brazilnuts.

Since then the Gavioes have had to contend with still more problems. Eletronorte is building a 500-kilovolt transmission line through Mãe Maria, which will mean the destruction of some 800 brazilnut trees as well as several thousand valuable hardwood timber trees. The negotiations for compensation involved Eletronorte, FUNAI, the central government, and the Gavioes. In 1978 the Indians asked for $560,000. Eletronorte found a legal loophole that appeared to rule out compensation payments to Indians. The company successfully argued that Brazilian law gives Indians use of the land and its assets but not ownership of them, which is reserved to the government. The Indians could use all the trees on their land, but the government, as landowner, has a right to remove some or all of the trees. In June 1977 the president of Brazil issued a decree giving Eletronorte access to the land in the interests of national development. The government would not pay compensation to the Gavioes. There it might have ended, but the Gavioes hired one of São Paulo's top lawyers and kept fighting. In 1980 Eletronorte agreed to pay the Gavioes compensation of $830,000.

Just as these negotiations ended, Brazil's state mining company, Companhia Vale do Rio Doce (CVRD), announced its plan to build an electrified railway line through the Mãe Maria Reserve. As many as twelve trains, each with 160 freight cars, will travel twelve miles through the reserve every day for the next eighty years. Track maintenance will require the building of access roads, workers' camps, and sand and gravel pits—all of which will cause disruption and pollution in the reserve. The Gavioes' objections were overruled, and their demand for $230,000 in compensation has been ignored. CVRD won with its arguments that rerouting the track away from the reserve would be too expensive and that the Gavioes, on the verge of extinction twenty years ago, are now so acculturated that they will not be disturbed by the trains.

. . .

The Amazon lends itself to the sort of statements that made Ripley famous. It has more than a thousand tributaries, several of which are larger than the Mississippi. Oceangoing vessels can cruise all the way to Manaus, 2,300 miles inland. One fifth of all the freshwater on earth flows every day through the Amazon, more than through the next eight biggest rivers on earth combined. Springs that are only 100 miles from the Pacific trickle eastward into rivulets, brooks, streams, and larger and larger rivers until finally, 4,000 miles from its source, the water pours into the Atlantic Ocean with such volume and force that 100 miles offshore a sailor can reach down to the ocean and scoop up a cup of freshwater. Marajó, an island in the Amazon estuary, is the size of Switzerland. At the river's mouth its north and south banks are farther apart than are Paris and London.

The energy potential of the rivers of the basin, leaving out the Amazon itself, is 100,000 megawatts. That makes the Amazon Basin the equivalent of an oil well that can produce five million barrels of oil a day and never run dry. Brazil is paying over $25 million a day to import just one million barrels a day, so it has a great incentive to develop its own hydroelectric reserves. Right now there are two small dams in the Brazilian rainforest with a combined output of only 65 megawatts. But by the year 2000, if Eletronorte's plans go according to schedule, the rivers of the Amazon Basin will be supplying 22,000 megawatts, 40 percent of the electricity Brazil will then be using.

The history of hydroelectric dams in tropical rainforests has not been a happy one. The first one, completed in 1964, created Lake Brokopondo in the former Dutch colony of Surinam, just north of Brazil. Five hundred and seventy square miles of dense virgin rainforest was flooded for this reservoir. As the trees decomposed, they produced hydrogen sulfide and a stink that brought complaints from people many miles downwind. For two years workers at the dam had to wear gas masks. The worst effect of the decomposing vegetation was that it made the water more acidic and corroded the dam's expensive cooling system. The cost of extra maintenance and of repairing and replacing damaged equipment was estimated in 1977 to have added up to $4 million, or over 7 percent of the total cost of the project.

A lake into which decaying trees are releasing nutrients is an attractive habitat for waterweeds. They appear first on the edges of the lake, fed by the nutrient-rich water lapping up and down on the shores. Eventually they break loose and float into the main part of the lake, forming dense mats and anchoring in the branches and on the trunks of still-standing trees. At Brokopondo, once the dam was built, water hyacinth, which up to then had been rare in the Suriname River, began to spread over the surface of the lake. Water hyacinth is an appealing plant with shiny green oval leaves and a beautifully delicate, pale violet flower, but colonies of it can cause tremendous damage and inconvenience in lakes and canals, impeding navigation and getting entangled in machinery. When the reservoir is low, they lie on the shore—a smelly, slimy, slippery obstacle to anyone wanting to get near the water. Within a year, a fifty-square-mile carpet of water hyacinth was floating on Lake Brokopondo, and by 1966 the weed covered almost half the reservoir. A floating fern called *Ceratopteris* covered another 170 square miles.

The second and newest dam in the Amazon Basin, Curua Una near Santarém, two hours by plane from Tucurui, began operating in 1977. Though its reservoir is only one twenty-fifth the size of Tucurui's, people forty miles away complained of the sulfurous stench for months after the uncleared reservoir began filling. Once the smell died down, the real troubles started. By 1982 the steel casings of both turbines had been corroded by the acidic water. Replacement casings made of stainless steel cost over $5 million.

Within a short time about half of Curua Una's lake was covered with floating mats of water hyacinth and sedge. One consequence was that many of the fish in the lake died. Water hyacinth is unpalatable to most fish—and even toxic to many. In addition, it absorbs the available nutrients and keeps the sun from penetrating into the depths of the lake. Other plants, and the fish that depend on them, have little chance of survival.

At Jupia, a dam on the Paranã River outside the Amazon region, the sheer weight of the mass of water hyacinths growing on the lake's surface has snapped steel cables. Special filters designed to protect the turbines from waterweeds have been so badly clogged that on occasion the dam has had to be shut down. But those are minor inconveniences compared with the fact that, floating on the water's surface or stranded on the shore, waterweeds provide food, oxygen, and breeding sites for the carriers of two of the most dangerous and unpleasant diseases known to man—malaria and schistosomiasis.

Malaria is already the most widespread debilitating disease in Amazonia. The incidence of malaria in Tucurui was serious before the influx of workers to build the dam. In 1976 one person out of every five examined was found to have the disease. The dam will greatly increase the number of suitable breeding sites for mosquitoes in the region. Each time the water stored in Tucurui Lake is sent through the turbines to produce electricity, 350 square miles of ideal mosquito breeding grounds will be exposed. If the period of low water in the reservoir is longer than two weeks, there will be time for an entire generation of mosquitoes to breed.

Schistosomiasis (or bilharzia, as it is also known) is caused by a parasitic flatworm and is transmitted to humans by a few genera of aquatic snails that thrive in slow-moving, weedy tropical waters. The parasite's larvae swim from the snail and penetrate the skin of people swimming, washing, or just standing in water nearby. Once in the bloodstream, the larvae spend time in the liver and then, depending on the species, take up residence around the bladder or intestines of the human host. Symptoms range from chronic diarrhea and urinary blood to more severe, potentially fatal ailments of the liver, kidneys, spinal cord, and other organs. Available treatments are costly and carry potential dangers of their own.

Fourteen million Brazilians have schistosomiasis. At present the Tocantins River itself does not harbor in any great numbers the snails that carry the parasitic flatworm. But Robert Goodland, a World Bank ecologist who undertook an environmental assessment of Tucurui for Eletronorte in 1977, believes that a few years after the lake is filled, it "will provide ideal conditions for snail proliferation." Since the host snails occur north, south, east, and west of Tucurui and since many of the dam workers are from the poor northeast of the country, where the disease is widespread, snails are bound to become more numerous in the area.

The government of Surinam spent $2.5 million spraying Lake Brokopondo in a largely successful campaign to destroy the weeds. The herbicide they chose was 2,4-D, one of the main components of Agent Orange, the defoliant used by the U.S. army in Vietnam and which many fear is linked to terrible birth defects in the children of American soldiers and Vietnamese civilians. Indians and other local people who were used to drinking, fishing, and canoeing in the waters of the Suriname River found their river first clogged, then poisoned. At Curua Una scientists have convinced the authorities that manatees are the answer to the waterweed problem. These gentle aquatic mammals, sometimes known as sea cows, feed on water hyacinth. Thirty manatees have been imported into Curua Una, but at least twice that many will be needed to keep up with the proliferating plants.

The rate at which the trees and other vegetation flooded by a reservoir decompose is crucial to the quality of the water, and the main determinant of that rate is the amount of free oxygen available. If the lake has a small surface area and is deep and stagnant, the decomposition will be anaerobic—that is, it will occur in the absence of free oxygen. Under these conditions, it is likely to go slowly and take many, many years. Among the gases that are produced as a by-product of anaerobic decomposition are methane, which is inflammable; hydrogen, which is explosive; hydrogen sulfide, which is poisonous and produces the "rotten-eggs" smell; and carbon dioxide, which increases the acidity of the water. At Brokopondo, where most of the lake experienced anaerobic decomposition, the lack of oxygen in the reservoir water had repercussions more than fifty miles below the dam. Here fish were found to be stunned by oxygen-deficient water that had passed through the dam's turbines.

Aerobic decomposition occurs in large lakes that are shallow enough for sunlight to penetrate, creating oxygen by photosynthesis in the underwater vegetation, and turbulent enough for the oxygen to be circulated throughout. Most of the Tucurui Reservoir is likely to undergo aerobic decomposition. This means that the forest will break down faster and that the effects of decomposition will be felt more immediately and more severely. But it also means that things should return to normal more quickly.

The breakdown has already begun, heralded by a putrid smell, in the small sections of forest that Eletronorte has flooded early for experimental purposes. These areas, and on a larger scale the whole of Curua Una, present an eerie, ghostly landscape. Many thousands of leafless, barkless skeletons of trees stick out of the lake, a haunting and strangely beautiful sight.

At Tucurui the area to be flooded contains something like twenty-six million cubic yards of good-quality timber, worth many millions of dollars. There are some very valuable species, such as mahogany and brazilnut trees. The rest could be converted to pulp or used in the wood-to-methane fuel project the government has scheduled for the Tucurui region. But Eletronorte will not clear the area first. As has so often been the case in Amazonian development schemes, a large part of the forest will be destroyed and the trees simply wasted.

Eletronorte argues that clearing the entire area of the reservoir is unnecessary and impracticable. "There has been a great claim that it should be cleared," said Barbosa. "But there was never a plan to do that. It would be absolutely uneconomic. If it was a question of having a great ecological disaster if the trees were not cleared, then we would have to abandon the dam, because there is no way we could afford to clear it. We just don't have the time. We know there will be problems in environmental terms, many serious problems, but it is a matter of economics. There won't be any complete disaster, and what we cannot solve, well, that's the price we have to pay. That's not just my view; it's also the view of Eletronorte."

On the schedule set for the completion of Tucurui, originally five years from start to finish, there certainly was not enough time to clear the lake area. Assuming the existence of a fair-sized pulp mill with a daily output of 750 tons, it would take approximately 115 years to utilize the wood from eight hundred square miles. There are only a handful of pulp mills in the world that can deal with the variety of timbers obtained from clear-cutting a tropical rainforest, and none of them is in Brazil. Other, less ambitious, suggestions for salvaging some of the reservoir forest included setting up small portable or floating sawmills so that the wood could be processed for local use, and continuing to salvage wood even after the reservoir was filled. In his report to Eletronorte, Goodland pointed out that being in a lake can actually increase the value of many trees. The water strips the trunks of their bark and branches and is a convenient way to transport them.

Originally Eletronorte did not intend to clear any of the forest that was to be flooded. But after Goodland's report, with its warnings about the risks of disease, acidity, waterweeds, and fish deaths, the company decided to clear selected areas. The clearance of twenty-five thousand acres was deemed vital to the proper functioning of the turbines. Several years were lost while the company decided to whom to award the contract. Finally it was given to CAPEMI, a diversified company with interests in everything from insurance to cattle ranching. In fact one of the few things in which CAPEMI had no interest or experience was logging. CAPEMI is better known in Brazil as the company that handles the army's pension fund. With the help of French consultants and a $25-million loan from the Banque Nationale de Paris, CAPEMI set to work, or rather failed to. In early 1983, having cleared only a fraction of the area it was supposed to, CAPEMI went bankrupt, owing more than $3 million to its three thousand workers. Eletronorte began desperately searching for a way to do the crucial clearing before the reservoir is flooded. There is probably not time to extract and sell the trees. The best Eletronorte can hope for is that it will be able to fell trees in the key areas and cover them with earth to accelerate their decomposition. Some company officials believe that the only way to get the job done in time is to fell the trees chemically—by spraying defoliants from airplanes. Public opposition shelved, but did not kill, that plan.

Eletronorte still has no clear idea of what will happen to Tucurui once the forest is flooded. Nor has anyone else. Many ecologists who have carried out studies in the area are worried. "In ecological terms," one expert on fish told me, "it's a disaster. The dam will suddenly change the condition of a great area. We have found more than two hundred species of fish here, and about 10 or 20 percent of them are new. Many of them have never been seen in any other river. Some species will certainly disappear from the river after the dam."

More than ten thousand animals were rescued from Brokopondo as the reservoir was flooded—including sloths, monkeys, deer, tortoises, and porcupines. There has not been a proper biological inventory of Tucurui, but it is rich in wildlife. It was here, for example, while he was traveling with Alfred Wallace, that the Victorian naturalist-explorer Henry Bates first saw a trogon, a beautiful bird whose rose-colored breast and glossy green back make it one of the most striking inhabitants of the rainforest. Eletronorte did not plan to mount an animal rescue. However, pictures in the Brazilian press of animals drowned or clinging to treetops after the filling of the reservoir of Itaipu Dam on the border with Paraguay may make it change its mind.

Officials are very concerned about the possibility that ecological problems could cause expensive damage to equipment. They sincerely hope there will not be a repeat of the Brokopondo experience. They trust that the ordeal of decomposition will come and go more quickly than at Brokopondo because the Tocantins River is faster flowing than the Suriname. It takes ten years for the Suriname River to fill the 3-billion-gallon Brokopondo reservoir. At Lake Tucurui, where the water will change 6.7 times a year, the acidic water should be flushed out of the system much faster. The test will begin when the dam closes and the reservoir starts filling, an often-postponed event that is now scheduled for 1986. Meanwhile, an official of Eletronorte's parent company, Eletrobras, confessed, "Tucurui has us scared to death."

. . .

I went with Barbosa and João Basilio, a biologist working for Eletronorte, to look at an experimental area of flooded forest. We drove for twenty-five miles through the forest along a newly cut dirt road west of the reservoir. The economic and technical arguments against clearing the reservoir area seemed not to apply here, just a few hundred yards outside the area to be flooded. On either side of the road stretching to the horizon are great, treeless cattle ranches ranging, I was told, from 125 to 25,000 acres. Most of them are on land set aside by the government to protect the watershed of the dam. The rare patches of forest left standing were in areas that are destined to be flooded.

The road winds through the forest, a ten-yard-wide gash of freshly exposed red earth. Where it cuts through hills, huge mounds of excess earth are piled alongside the road. Where the land is flat, the bared soil stretches on either side for many yards before meeting scraggly low scrub, or barbed-wire-enclosed pasture. Though the surface of the road is smooth and well leveled, the countryside is far from being flat. I had an idea that the Amazon was a vast plain, modified only by the most gentle dips and rises. In fact, traveling at speed along that road was terrifying. Luckily we were in a Volkswagen. Nothing bigger, it seemed to me at the time, could have fit into the hairpin bends between the foot of one hill and the start of the next. Once or twice I was quite sure that we hadn't actually touched bottom. The two men laughed about the illusion of "the Great Amazonian Plains," a description of the rainforest that every Brazilian schoolchild learns.

Deforestation of the surrounding uplands and the slopes of the river basin (as opposed to the clearing of the area to be flooded) threatens the smooth functioning of the dam. Accelerated erosion could lead to increased siltation of the reservoir and its feeder rivers. And, because rainwater rushes off denuded lands rather than being absorbed and gradually percolating through to the river system, deforestation leads to exaggerated wet and dry seasons. The result could be a shortage of water to operate the turbines at full capacity during the dry season.

The prospect of a multimillion-dollar investment being destroyed by siltation caused by deforestation may seem fanciful, but it has already happened more than once. The Anchicaya Reservoir in Colombia silted up almost completely within fifteen years of its inauguration, requiring the construction of a second multimillion-dollar dam farther up the Anchicaya River. In the Philippines, the reservoir of Ambuklao Dam has silted up so quickly as the forests of the Agno River watershed have been cut down that it is no longer fully operational. It was planned to last seventy-five years, and to pay back its costs in sixty-two years, but is now expected to last only until its thirtieth year in 1985, by which time, according to the U.S. Agency for International Development, losses due to siltation will have amounted to more than $25 million. Intensive agriculture in the watershed of the Rio Alto Pardo in Brazil caused such severe runoff that in 1977 two dams on the river were completely smashed.

Reservoirs hold water in either "dead" or "live" storage. Water in live storage is available to turn the turbines and generate electricity. Dead storage water simply fills up the depressions in the riverbed, without adding to the volume of water available for power generation. Dead storage areas provide a buffer zone for sediment, since as far as the dam's performance goes, they may as well be filled by sediment as by water. Many dams have much more dead than live storage. In Tucurui's case slightly more than half the total capacity of 56 million cubic yards is live storage. In some months more than 105,000 cubic yards of earth washes into the Salto Grande Dam on the Santo Antônio River. At that rate Tucurui's dead storage would be exhausted within twenty years, at about the time that the dam is calculated to start turning a profit.

Eletronorte is already engaged in a battle against erosion on the banks of the as-yet-unfilled reservoir. When I was there in 1982, some of the banks had been planted with grass before the previous rainy season. Those that had not been were marked with deep gullies carved out by the rain as it carried the denuded soil into the lake bed. Planting grass the way Eletronorte goes about it is a tedious and expensive process. At first they seeded the banks in the normal manner, but that method required a layer of topsoil, which the first rains carried away. So now the grass is germinated in a nursery and then planted out blade by blade, one at a time.

Back in 1977 Goodland wrote that "the acceleration of deforestation in the Tocantins watershed amazes even SUDAM." Eletronorte has no figures on how much deforestation is now taking place, even on the surrounding federal "public utility" land. Company officials say it's not their business to control deforestation of the watershed. "We can't tell people not to cut down trees," Barbosa told me. "All we can do is try to influence the people and agencies who do make policy."

To protect its $4-billion investment from the effects of siltation, the government will have to control deforestation not only in the immediate area of the dam but all through the 146,000-square-mile watershed. Brazil has an impressive roster of laws designed to protect the environment, but they are not well enforced and most are riddled with loopholes. One of the best-known environmental laws, for example, is the so-called 50 percent rule forbidding people who buy forestland to clear more than half of it. The law does not, however, prevent the owner from selling the uncleared half to someone else, who can clear half of it and sell the rest, and so on down the line. Brazil's forestry code also requires large-scale rainforest operations to reforest a certain amount of land in compensation for the land they have degraded. But the plantation need not be on the already degraded land; instead (and much more profitably) the company can abandon that land and establish its plantation on a completely new area, felling or burning virgin forest to do so, without incurring any further obligations.

. . .

The main purpose of the Tucurui Dam, which will ultimately produce 8,000 megawatts of electricity, equal to six Three Mile Islands, is to encourage industrial development of the Amazon by providing an abundant and cheap supply of electricity. One of Tucurui's biggest customers will be the electric railway through the Gavioes' Mãe Maria Reserve. The railway is part of the biggest enterprise ever attempted in the Amazon, the Carajás iron ore project. The Carajás Mountains, only a few hundred miles southwest of Tucurui, contain the world's largest known deposit of iron ore. The railway will carry the ore 550 miles northeast to a new deep-water port at São Luís on the Atlantic coast. The $5-billion project is being financed by Brazil and a number of foreign partners—including four Japanese financial groups, which hold 10 percent; several American banks, with 5 percent; and the World Bank and the EEC, which are putting up 7 and 10 percent, respectively, of the cost. In return for the EEC loan, European steelmakers are guaranteed supplies of iron ore at what community officials call "favorable prices" and a Brazilian opposition politician called "banana prices." At the urging of the World Bank, CVRD has employed three ecological advisors. Also at the bank's instigation, the company has offered FUNAI $13.6 million over five years to be used on behalf of the Indians. Nestor Yost, the executive secretary of Carajás, has suggested that the agency use the money to help the Indians "reach a level of acculturation to the point of being assimilated as workers into the programme."

Another big user will be the $2.6-billion joint Japanese-Brazilian aluminum smelter and refinery near Belém. This huge complex, which is scheduled by the end of the century to consume up to 5 percent of Brazil's electricity supply, will process bauxite from a Canadian mine on the Trombetas River and from another in Paragominas owned by Rio Tinto Zinc. Royal Dutch Shell and the U.S.-owned Alcoa Alumínio also have a bauxite mine on the Trombetas River. Their $1-billion refinery and smelter at São Luís will buy electricity from Tucurui. Some of Tucurui's electricity will also be supplied to Belém and, by linking into the national grid, to the Brazilian Northeast, but the emphasis is on building up the Amazon as an industrial region. Foreign investment is crucial to this plan. The bulk of the industrial production will be for export because Brazil—which has the biggest balance-of-payments deficit in the world, over $90 billion—badly needs foreign exchange.

Though the Carajás iron ore scheme is the biggest project ever undertaken in the Amazon, it is just a small part of a much more grandiose scheme. The Grande Carajás, as the superproject is called, will extend over an area of forest the size of France and Great Britain combined. It covers 300,000 square miles, one sixth of Amazônia Legal, the government-defined region that goes beyond the ecological limits of the Amazon Basin. The area includes more than twelve thousand Indians and more than twenty Indian reserves, yet CVRD's assessor of mines acknowledges only the "possibility of Indian nations within the GCP area." Grande Carajás is in the grand tradition of spectacular assaults on the Amazon, but if it ever gets off the ground, it will dwarf all its predecessors. It will include up to thirty other mines, as well as metal processing, hydroelectricity, forestry, ranching, and farming projects. Ten new cities will be built for the workers and migrants who will soon flood into the region. One of the sub-schemes promoted by the executive secretary of the Interministerial Council of the Grande Carajás Program calls for clear-cutting ten thousand square miles of forest in order to establish plantations twenty-four times bigger than those at the Jarí project, the $1-billion failure not far away that Daniel Ludwig, said to be the richest man in the world, abandoned in 1982. The $62-billion cost of Grande Carajás is to come largely from foreign businesses and development agencies and private Brazilian companies. Cheap electricity from Tucurui is one of the great attractions to potential investors.

One of the strongest arguments in support of Tucurui is that its electricity will reduce Brazil's annual $9-billion oil import bill. By providing the equivalent of 400,000 barrels of oil per day, Tucurui should save the nation $12 million a day, or $4.38 billion a year (assuming a cost of $30 per barrel). But there is a catch. In order to obtain international finance for the iron ore and bauxite projects, the Brazilian government agreed to give foreign investors big tax breaks, generous import quotas, and electricity at concessionary prices averaging 30 percent below the market price. Ordinary consumers and local industrialists, the Brazilian taxpayers who paid for the dam to be built, will pay market prices for any electricity they may receive from Tucurui.

Another hitch is that Brazil already has a nearly 25 percent surplus of generating capacity. Brazil's electricity demand fell by 10 percent in 1981, instead of rising, as government analysts had forecast, by 14 percent. Brazil had the generating capacity to produce 154 million megawatt hours of electricity, but could sell only 124 million. In any case, most of Tucurui's production will not go to replace electricity produced by imported oil for the heavily populated South. Tucurui will be largely supplying, at a discount rate, new industrial customers in the North. Nonetheless, Eletronorte believes, as Barbosa told me, that in the Amazon "any dam is economic, most of all if you consider that the land is free. The only price is the environmental one."

. . .

The plans for hydroelectric power from the Tocantins do not stop with Tucurui. A series of eight large dams, of which Tucurui is the first, and nineteen smaller ones, is planned for the Tocantins and twelve tributaries. Eletronorte estimates the electrical potential of this project at 22,000 megawatts. This is more electricity than the whole of Brazil now consumes. The Tocantins River Basin Hydroelectric Project, as it is known, will convert the river into an almost continuous chain of lakes 1,200 miles long, stretching from Tucurui to within 75 miles of Brasília.

The irony of Brazil's current race to turn the Amazon into a series of dams and reservoirs is in its resemblance to the South American Great Lakes scheme thought up by Robert Panero and promoted in the 1960s and '70s by the Hudson Institute, the right-wing think tank run by the late futurologist Herman Kahn. Panero's idea was simple. Dam the Amazon River near Santarém and create a reservoir that would flood more than forty thousand square miles. It would be bigger than the biggest lake in the world. That dam would produce more than fifty times as much energy as the Aswan High Dam, but there would also be forty or so others, creating electricity for sale to North America, draining potential farmland, and creating a great navigable chain of inland waterways. The Amazon Basin countries—Brazil, Colombia, Guyana, French Guiana, Peru, Surinam, and Venezuela—would, according to Panero, have to accept a certain blurring of their national borders. They responded badly to his vision. They suspected they were being set up. Various theories about the real purpose of Panero's scheme were floated. Was the plan to ship American black power agitators back to the jungle? Or was it a plot to turn the Amazon into a bomb shelter for the right people in the coming nuclear war? Even if it was none of these dramatic things, even if it was just what it looked like, another Yanqui self-enrichment scheme at the expense of the South Americans, it was not welcome. The Brazilians had long been nervous about their undefended northern frontier. Panero's crazy scheme was one more reason to secure their borders against foreign infiltration. And the best way to do that was to develop the region, to stake a claim and get people out there. The Trans-Amazon Highway helped, but officials were soon frustrated by the repeated failures of small-scale agricultural projects on unsuitable rainforest soils. Industrial development seemed a better bet, and hydroelectricity was the best way to start on it. Soon Brazil will have just the sort of Great Lakes system Panero proposed, but at least it will be self-inflicted.

The lessons learned at Tucurui should be invaluable to the Brazilians, because building hydroelectric dams in the rainforest promises to be a boom industry well into the next century. The only question is whether, on the punishing schedule Brazil has set itself, there will be time to learn from past experience. The speed and scale of Brazil's hydroelectric program owes something to the intense desire of its technocrats to shake off the image of a developing country and enter the big league of industrialized nations. Whether or not the electricity produced by a series of Amazonian dams is necessary and economical, the sheer magnitude of the engineering feat gives many Brazilians a sense of national pride.

Goodland's pioneering analysis of Tucurui led Eletronorte to look at several problems that the company had not previously taken seriously, including the problems associated with forest decomposition and the question of the dam's effect on Indian lands. But a study that may recommend changes in the design or execution of a billion-dollar project is more likely to be heeded if its findings can be studied before construction has begun. So far this has not happened in Eletronorte's hydroelectric program. In 1980, partly as a result of Goodland's prodding, three years after construction on the dam had actually begun, Eletronorte finally commissioned a more detailed environmental study at Tucurui, covering such subjects as disease vectors, water quality, forest decomposition, fisheries, and waterweeds. "Whatever we find," one scientist working at the dam told me, "there won't be time to alter the design of the dam in any significant way."

Eletronorte also hopes to build nine or ten dams on the Xingu River, another tributary of the Amazon. The first two dams likely to be built on the Xingu will have a combined output of 14,000 megawatts, almost twice that of Tucurui. Several thousand Indians, and many more non-Indians, are likely to lose their lands and their livelihoods no matter which of the several proposed schemes goes ahead. One plan would flood 2,300 square miles, including Altamira, the third largest city in the Amazon. The overall program was announced in 1978. FUNAI has still not examined how it will affect the Indians in the region.

In addition to these two huge river basin projects, there are at least three other Amazonian hydro projects now underway or planned. The most controversial of these is Balbina Dam, a joint Brazilian-French venture designed to supply power to the fast-growing industrial free trade zone of nearby Manaus. Although Balbina's reservoir will be the same size as Tucurui Lake, its electrical output will be less than 4 percent of Tucurui's. Electricity produced at Balbina will be more than four times as expensive as the same amount of electricity from Tucurui. Parts of Balbina are, like Tucurui, scheduled to be cleared before the reservoir starts filling in 1986, but decomposition is expected to be a very serious problem at Balbina. While assuring me that Tucurui would not suffer from the effects of forest decomposition, an Eletronorte official told me, "I am extremely worried about Balbina. I don't talk about it. There are going to be serious problems there." Construction began at Balbina in 1981; environmental studies started in late 1982. Balbina will also flood a large part of the Waimiri-Atroari Indian Reserve. So far the two thousand Waimiri-Atroari living there have not been consulted about the reservoir, nor, apparently, has FUNAI made any plans to relocate them.

Eletronorte has identified seventeen other possible sites for hydroelectric dams within 325 miles to the northeast of Manaus. One of these is Cachoeira Porteira on the Trombetas River, one of the Amazon's tributaries from the north. Construction is due to begin in 1984. La Porteira's reservoir will flood 460 square miles of a reserve set aside for the Wai-Wai Indians. (Many Wai-Wai are Seventh Day Adventists and speak fluent French, English, and Portuguese as a result of contact with different missionaries over the years.) "The Indians are as happy as can be," one Eletronorte official assured me. "We've given them even better land to the north. It's pretty much what they wanted. They're relieved that they're getting any compensation at all." The reservoir will also flood part of the Trombetas Biological Reserve. Brazil's National Parks and Biological Reserves are the responsibility of the Forestry Department's National Parks section, headed until recently by the indomitable Maria Tereza Jorge de Padua, a woman of ruthless charm and tenacity. She was not happy about the prospect of one of her hard-won reserves being put under water. "We bite into a few acres of Trombetas and Maria Tereza is kicking like hell," one planner complained to me. Maria Tereza knew that she was fighting a losing battle, though she gave her opponents a few white hairs in the process. "They say yes to the creation of a protected area and afterward they start doing things there," she told me. "I can stop them for a while but not forever."

The site chosen for the Cachoeira Porteira Dam is along a beautiful stretch of the river. I visited it with a group of engineers from Eletrobras. We all stood in silence looking at the river, admiring the view, when one of the men turned to me and smiled. "We're going to save all this for posterity," he said. "We're going to cover it up with water so that no one can disturb it."

2
In the Beginning

"Its lands are high; there are in it very many sierras and very lofty mountains. . . . All are most beautiful, of a thousand shapes; all are accessible and filled with trees of a thousand kinds and tall, so that they seem to touch the sky. I am told that they never lose their foliage, and this I can believe, for I saw them as green and lovely as they are in Spain in May, and some of them were flowering, some bearing fruit, and some at another stage, according to their nature." Christopher Columbus's description to Ferdinand and Isabella of the forests of Hispaniola (the island that is now divided into Haiti and the Dominican Republic) is the first written account of a rainforest. It contains the seeds of two different, but not incompatible, views of the rainforest, the romantic and the scientific, both of which still have their adherents.

For the four centuries after Columbus's discoveries, travelers and scientists had no special name for tropical rainforests. They contented themselves with calling them forests, or tropical forests. Alfred Wallace in his 1878 book *Tropical Nature* was so precise as to call them "the primeval forests of the equatorial zone." In 1898 the German botanist A. F. W. Schimper coined the phrase *tropische Regenwald* (tropical rainforest).

Schimper defined such a forest as "evergreen, hygrophilous in character [that is, growing in wet places], at least one hundred feet high, but usually much taller, rich in thick-stemmed lianes and in woody as well as herbaceous epiphytes." Botanists now distinguish, or try to, between as many as thirty or forty different types of rainforest, including, for example, evergreen lowland forest, evergreen mountain forest (subdivided into broad-leaved and needle-leaved), tropical evergreen alluvial forests (further divided according to the degree of flooding), semi-deciduous forests (broken down into lowland and mountain forests), and so on. Because nature is continuous and science seeks clear-cut categories, these definitions are rarely wholly satisfactory, and each brave attempt to impose order on a complex and poorly understood ecosystem spawns revisions and adjustments and new tries.

But for the purposes of a general survey, we can, like Schimper—and Wallace, Bates, and Darwin before him—make do with a fairly simple classification. The defining characteristics of tropical rainforests are temperature and rainfall. At one end of the spectrum are forests with high rainfall (160 to 400 inches a year) and a high average temperature (80°F.), and without pronounced cold or dry spells. These are the equatorial evergreen rainforests. Within this category differences in soils and altitudes give rise to many subcategories, such as evergreen montane forest, swamp forest, and lowland rainforest.

Moving away from the equator, north or south, the climate gradually becomes more seasonal, with increasingly pronounced cold and dry spells. The second main group of tropical rainforests (more properly called tropical moist forests, or tropical semi-deciduous forests) are marked by this seasonality. They get less rain (40 to 160 inches a year) and have a less constant temperature, and a marked dry season, when many or all of the trees lose their leaves. These seasonal forests are not as rich in species as the warmer, wetter equatorial forests. Because for at least part of the year their canopies are more open to the sun than equatorial rainforests, they have a more luxuriant understory. They too can be further divided according to their soils and altitude, as well as finer gradations of temperature, rainfall, and seasonality.

Two thirds of the world's rainforest is the wetter, richer, equatorial type. The major concentrations are in lowland Amazonia, the Congo Basin, Sumatra, Borneo, and Papua New Guinea. Burma, Thailand, Cambodia, Java and Sulawesi, northeast Australia, and parts of West Africa and South America all have important areas of seasonal rainforest.

Several hundred million years ago all the future continents huddled together in one great landmass scientists now call Pangaea. About 210 million years ago Pangaea separated into two supercontinents, Gondwanaland (Africa, South America, Australia, and Antarctica) and Laurasia (North America, Europe, and Asia). At that time dinosaurs held sway on land, crocodiles and ichthyosaurs ruled the sea, and pterodactyls dominated the skies. Primitive palmlike plants called cycads were the main vegetation. Gondwanaland, the southern supercontinent, began to break up some 180 million years ago, and the continents finally settled in more or less their present positions about 3 million years ago.

Meanwhile there had been a revolution in life on earth. Around 125 million years ago, the first flowering plants suddenly emerged. They probably first appeared somewhere between India and Fiji, a part of the globe that today has the richest plant life of anywhere. Mammals, well adapted to this new form of plant life, increased in number and diversity. Dinosaurs and other reptiles, however, continued to predominate until 65 million years ago, at the end of the Cretaceous period, when, for reasons scientists can't agree on, they suddenly (on a geological timescale) became extinct. Many of the families of animals and plants that make up tropical rainforests existed in the Cretaceous period, while the continents were still relatively near one another.

Forty-five million years ago tropical rainforests extended across much of the world. Scientists have found fossilized pollen grains of tropical rainforest species in London, Tennessee, and Alaska. It may be that the flora of the drier tropical and temperate regions has arisen relatively recently from the ancient rainforest flora. From twenty million to five million years ago, changes in climate and topography pushed the rainforest to lower latitudes.

The rainforests of the New and Old Worlds are similar in structure and appearance, but they have hardly any species in common. Many genera, however, and even more families occur in more than one area. The plant and animal links between the three main rainforest areas may be explained by their shared continental past, or may be the result of long-distance dispersal. Of the 161 families of flowering plants that grow in the Amazon rainforest, all but 21 are also found on other continents. Most Amazonian genera are also found in other rainforests, but the unevenness of the distribution is fascinating. The genus *Hirtella,* for example, grows only in the Amazon basin and East Africa. Another rainforest genus, *Symphonia,* has 17 species, of which one, *S. globulifera,* occurs in Central and South America, the West Indies, and West Africa, and the other 16 exist only in Madagascar. The pineapple family (Bromeliaceae), which has 1,600 species in 60 genera, is confined to South America, except for one species, *Pitcairnia feliciana,* which grows also in the rainforests of Guinea in West Africa. Tapirs, a rainforest relative of the hippopotamus, originated in the northern supercontinent, Laurasia, when it had a tropical climate. Now tapirs are found only in two widely separated areas—South America, on the one hand, and Malaysia and Sumatra, on the other. There are only 3 species of lungfish in the world: One lives in South America, one in West Africa, and one in Australia.

The interchange of species between islands in the Pacific led Alfred Wallace to one of his great insights. Wallace owed his revelation to an inconvenient shipping schedule. As he remarks in his travel journal, "Had I been able to obtain a passage direct to Macassar from Singapore, I should probably have missed some of the most important discoveries of my whole expedition to the East." Instead he was forced to call in at Bali in June 1856 to wait for a ship that would take him to Macassar. During his two-month wait, Wallace concluded that although Bali is only fifteen miles from its neighboring island, Lombok, the two islands "are the extreme points of the two great zoological divisions of the Eastern Hemisphere." The invisible barrier between Bali and Lombok, now called Wallace's Line, marks, as he wrote to his friend, the naturalist Henry Bates, "two distinct faunas, rigidly circumscribed, which differ as much as those of South America and Africa. . . . Yet there is nothing on the map or on the face of the islands to mark their limits." In the first, notes Wallace, "the forests abound in monkeys of many kinds, wild cats, deer, civets and otters, and numerous varieties of squirrels are constantly met with. In the latter none of these occur." Bali, Java, Borneo, Sumatra, and Indochina are all linked by a single continental shelf, called the Sunda Shelf, whereas Lombok and the islands to the east, including New Guinea and Australia, originally part of Gondwanaland, are connected by the Sahul Shelf. Wallace's line marks the boundary between Gondwanaland and Laurasia, a relic of Australia's collision with Asia fifteen million years ago.

During the last two million years—with the continents settled and rainforests occupying more or less their present range—the world has been subject to a series of ice ages, alternating with wetter, warmer interglacial periods. We are now in an interglacial phase, but there is no reason to believe that the cycle has ended. The world's climate dries out during ice ages because, with more water locked up in ice sheets, less evaporates into the atmosphere. During such periods rainforests appear to have shrunk back to a few patches where the climate was still suitable. Afterward, as the ice caps melted and the climate became warmer and wetter, forests expanded again. This process of retreat and expansion seems to have occurred a number of times during the glaciations of the Pleistocene epoch, in Amazonia, in Australia, and in West Africa (studies using carbon dating suggest that large areas of the Congo rainforest were desert during the Pleistocene ice ages), with important effects on the evolution of the plants and animals of the forest.

The idea that the forest retreated to a number of "refuges" during past ice ages was first proposed in 1969 by a West German ornithologist, Jürgen Haffer. He identified a dozen places in the Amazon Basin that are particularly rich in bird species and have a high number of endemic species—that is, species that occur only in one area. Jürgen postulated that during the cool and dry ice ages, these were forested "islands" in a savanna sea, that acted as refuges for forest species. Plants and animals in each patch evolved differently in response to local conditions. The expansion of the forests at the end of each ice age gave plants and animals a chance to spread, retrenching during each new ice age in different combinations, each with different evolutionary pressures. Since then other researchers have found similar and sometimes overlapping "refugia" for a few other groups of plants and animals, including butterflies and lizards, both in South America and West Africa. This theory—still in an early stage of development—offers one explanation for the high degree of endemism and species diversity in rainforests, and one way of identifying those areas that would be most valuable to protect.

There is no evidence that the Malesian rainforests (Malesia refers to the islands of New Guinea, Borneo, Sulawesi, Java, and Sumatra, and the Malay Peninsula) were reduced to isolated patches during the ice ages. Nonetheless they are as rich in species and in endemic plants and animals as the "elite" refugia of the West African and South American rainforests. Scientists speculate that the contraction and expansion that encourages speciation took place in the Malesian rainforests in a different way. During glacial periods, sea levels dropped by as much as 600 feet. A drop of only 330 feet exposes land links between islands in Malesia. Scientists believe that large forests existed on land that is now below water and that those forests were drowned during interglacial periods, such as we are now experiencing. When water levels were low, species could migrate across the region; when they rose, the Malesian forest was literally reduced to islands and new species evolved in response to local conditions.

. . .

Throughout the wet tropics, rainforests are the natural vegetation. All along the equator, between the Tropic of Cancer and the Tropic of Capricorn, wherever the temperature is high enough and the rain heavy and regular enough, there is—or once was—rainforest. A few thousand years ago the rainforest belt covered 5 billion acres—14 percent of the earth's land surface. Man has already destroyed half that. Most of the damage has been done in the last two hundred years, especially since the end of the Second World War. Now Latin America has 57 percent of the remaining rainforest. Southeast Asia and the Pacific islands have 25 percent. And West Africa has 18 percent. About thirty-seven countries have significant areas of tropical rainforest, though three have more than half the total. Brazil, with one third, has by far the biggest share. Zaire and Indonesia have one tenth each.

Tropical rainforests are being destroyed faster than any other natural community. A United Nations study from 1976 offers the most optimistic assessment of forest loss. It found that, of the 2.4 billion acres of rainforest left in the world, 14 million are completely and permanently destroyed each year. That is almost 30 acres every minute of every day. In 1980 the U.S. National Academy of Sciences announced an even worse figure. It said that over 50 million acres of rainforest—an area the size of England, Scotland, and Wales—are destroyed or seriously degraded each year. The most comprehensive study to date, published in 1981 by the Food and Agriculture Organization of the United Nations, says that at present rates almost one fifth of the world's remaining tropical rainforest will be completely destroyed or severely degraded by the end of the century. According to that survey, if current rates of deforestation do not increase, Indonesia will lose 10 percent of the forest it had remaining in 1981 by the year 2000. The Philippines will lose 20 percent, Malaysia will lose 24 percent, and Thailand will lose 60 percent by 2000. In Africa, Nigeria and the Ivory Coast will be completely deforested by 2000, Guinea will lose one third, Madagascar 30 percent, and Ghana 26 percent of their few remaining rainforests. In Latin America, Costa Rica will lose 80 percent of its 1981 rainforest. Honduras, Nicaragua, and Ecuador will lose more than half; and Guatemala, Colombia, and Mexico will lose one third of their last rainforests by the end of the century.

Mainland India, Bangladesh, Haiti, and Sri Lanka have already lost all their primary rainforests. By 1990 the lowland rainforests—the richest in terms of plant and animal species—of peninsular Malaysia, Thailand, the Philippines, Guatemala, Panama, Sierra Leone, and the Ivory Coast will have been reduced to a few remnant patches. And, according to the 1978 United Nations *State of Knowledge* report, "at the present rate of forest destruction all accessible tropical forests will have disappeared by the end of this century." But some large blocks, in Amazonia and Zaire, for example, may survive—if no new roads are built to open them to loggers, miners, and settlers. At present rates Brazil will lose 8 percent of its existing forest by the year 2000, which, though it may sound small, is sixty-three million acres, two and a half times the size of Portugal. But the rate of deforestation in the world's remaining rainforests is likely to increase. In Brazil, for example, satellite studies carried out by the Brazilian Forestry Development Institute show that 60 percent of the area deforested by 1978 had been cut between 1975 and 1978. The pressures to open these areas will intensify sharply in the near future. Already the logging industry is turning its attention from the near-depleted forests of Southeast Asia to the relatively untouched riches of Amazonia. And vast industrial developments are underway or being planned in many of the few remaining unspoilt areas.

Why are these forests—the richest, oldest, most complex ecosystems on earth—being cut down at such a rate? Why destroy a forest? To sell its timber, to get at the gold and iron underneath, to get more land for agriculture. There are psychological motives too: the wish to conquer nature, the fear of the unknown, nationalistic and strategic desires to occupy uncontrolled regions.

Overpopulation is usually cited as the main cause of deforestation. Rainforests are often used as safety valves by governments to avoid pressure for land reform. The safety valve theory, however, is misguided. Rainforests are not empty. Small groups of people are already living wherever the forest can support human life. Nor is the intact forest idle. It conditions the soil, regulates rainfall, and maintains the water cycle far beyond the borders of the forest itself. Most attempts to turn forest into farmland have failed disastrously, damaging the forest, disrupting the soil and water balance for other farmers, and leaving the settlers even more desperate for land.

The true cause of agricultural settlement in rainforests is often iniquitous land distribution rather than simple overpopulation. Among the rainforest countries, only Haiti, India, and the Philippines have a population density higher than 400 people per square mile; Italy, Japan, Great Britain, Belgium, the Netherlands, and West Germany all have more than 500. Brazil, which has a policy of moving settlers into the Amazon rainforest, does not need that land for agriculture. Leaving aside the Amazonian forest, it has the same population density as the United States, about 65 people per square mile. Western Europe averages more than 400 people per square mile. The Netherlands is prosperous with 857 people per square mile. Brazil has 2.3 acres of farmland per person, which is more than the United States, the world's greatest exporter of food. Taking potential farmland into account but still leaving aside Amazonia, each person in Brazil could have 10 acres. Instead, 4.5 percent of Brazil's landowners own 81 percent of the country's farmland, and 70 percent of rural households are landless.

The most important issue is not population pressure but how the land is distributed. Omitting the 85 percent of Java's families that have no land at all, 1 percent of the island's landowners own a third of the land. In India more than half the arable land is owned by the top 8 percent of the rural population. In El Salvador fewer than two thousand families own 40 percent of the land. Even when one considers farmers who rent land from others, the picture isn't much better. Typically in developing countries less than 10 percent of the rural population farms more than half the land. In Peru 1 percent of the population farms more than 80 percent of the arable land.

Common though it is for government officials, businessmen, and international aid agencies such as the United Nations and the World Bank to blame deforestation on masses of poor people searching for land, it is not the main cause of forest destruction in many areas. Land hunger is not even the prime motivation in many government-sponsored settlement schemes. Some of the largest ones—in Indonesia and Brazil, for example—are largely intended to secure national sovereignty by establishing a civilian presence in frontier regions. In the words of one member of the Brazilian junta, "When we are certain that every corner of the Amazon is inhabited by genuine Brazilians and not by Indians, only then will we be able to say that the Amazon is ours."

In Latin America cattle ranching for the export trade is the chief culprit in rainforest destruction. Government figures attribute 38 percent of all deforestation in the Brazilian Amazon between 1966 and 1975 to large-scale cattle ranches, 31 percent to agriculture, and 27 percent to highway construction. The government gave fiscal incentives to 90 percent of the ranches, and more than half the agricultural clearing was under a government-sponsored peasant colonization program. That program has now been wound down, in favor of investment in large-scale logging operations, hydroelectric dams, mines, and industrial developments.

Logging, mining, and other industrial activities do not result from population pressure. "Those countries in which current forest harvesting is of greatest concern (Indonesia, Brazil, Malaysia, and Colombia)," a 1982 U.S. State Department survey said, "also have relatively low population density." Industrial development and settlement, however, often go hand in hand because the roads enable settlers to reach previously inaccessible forests, and because settlers prefer to plant on the conveniently pre-cleared land loggers and others leave behind them.

In Southeast Asia, Oceania, and Africa, logging vies with peasant agriculture as the main cause of deforestation. According to Food and Agriculture Organization (FAO) figures, peasant agriculture in Indonesia affects only 500,000 acres of rainforest a year, one quarter of the area annually affected by logging. Some heavily logged countries are now reaching the point of no return. Since 1960 more than half of peninsular Malaysia's rainforests have been logged, and the official forecast of the Department of Forestry is that remaining forest resources will be exhausted by 1990 at the latest. One quarter of the Ivory Coast's foreign earnings come from timber, but with over one million acres being cut by loggers each year, there will be no timber left by 1985. Sixty percent of Africa's rainforest is in the Congo; the Congolese government has scheduled 68 percent of the country's rainforest to be logged.

In spite of claims that rainforests must be sacrificed to the betterment of the poor and landless, the effect of most rainforest exploitation is to redistribute wealth upward. The permanent, widely distributed benefits of the intact forest—the protection of wildlife, water catchments, and soil, and the provision of food, medicines, and building materials—are turned into immediate, short-term profits for a small group of investors and consumers.

. . .

One of the most endearing traits, to my mind, of that most gentle of the Victorian rainforest travelers, Henry Bates, was that he never called the forest a jungle. But to many people tropical rainforests are jungles. The word itself came originally from the Sanskrit word *jangala,* meaning desert. Later it came to mean scrubland, and later still, any wild place. The word *jungle* has an overburden of mystery and danger; it also implies a challenge: to conquer or be conquered. The primeval forest is not a hospitable place for man. Although some peoples have learned to live in the forest, most rainforests are sparsely populated for very good reasons. Man does well to respect the forest and keep a cautious distance.

The Pygmies of the Congo Basin are among the true forest peoples. They have lived there for many thousands of years, and to them the forest is Mother and Father. It protects them and gives them life. When things go wrong, the Pygmies say that the forest is sleeping, and they sing and dance so that it will awake and protect its children. But, as Colin Turnbull made clear in his study *Wayward Servants,* the non-Pygmy Bantu-speaking groups who in the last several centuries have been forced into the forest by competition for land hate and fear it. "They people the forest with evil spirits, and they fill their lives with magic, witchcraft and a belief in sorcery." Their first desire is to cut down the trees that harbor evil spirits and bloodthirsty animals.

Panama's military rulers have made *La conquista de la selva* (the conquest of the forest) their slogan. The phrase has psychological and political appeal; it unites a macho assertion of power over the unknown with the naming of a new frontier where displaced peasants can, without threat to the owners of large estates, find new land. The main way of staking a legal claim on unoccupied land is to improve it. Throughout Latin America, as well as much of Africa and Asia, the only legally recognized way to improve forestland is to cut down the trees. In 1940 the Brazilian president summed up this view of the forest: "To see Amazonia is the heart's desire of the youth of this country. To conquer the land, tame the waters, and subjugate the jungle, these have been our tasks. And in this centuries-old battle, we have won victory upon victory."

The variety, complexity, and purposefulness of rainforest creatures inspire many who have seen them, if only through books or television, with a sense of the power and diversity of life itself. But to others rainforests are jungles, physically and morally threatening until tamed. In trying to subdue the jungle, they are bringing back the original meaning of the word.

When investors destroy tropical forests to get timber, diamonds, hydroelectric power, iron, and grazing land for beef, and when poor people move into the forest searching for land on which they can grow food, we must ask questions. Will the forest be so disturbed it can no longer protect wildlife, soils, and watersheds? Will the forest be squeezed dry in a few years or can the farming, mining, or whatever go on forever? Who will benefit from the new use—the people of the forest? the poor of the country? a domestic elite? foreign consumers? Although we may not live in rainforest countries, we have a right to ask these questions because millions of our fellow creatures—humans, animals, and plants—depend directly upon the health of the rainforests; because hundreds of millions more, including those in the industrialized countries, rely upon them indirectly; and because so much rainforest is destroyed in our names and with our money.

3
Layers

The intrepid explorer of popular fiction who needs an ax to penetrate the tangled jungle is not in a pristine tropical rainforest. To be sure, travelers see a dense wall of green along roads and rivers, where the open edge of the forest is exposed to the sun. But deeper inside such a forest very little sunlight filters through the dense canopy, and the floor is surprisingly free of undergrowth. There are some low-growing plants and shrubs, and many seedlings and saplings, but it is generally easy enough to walk through. Except where the forest is interrupted by roads or streams, strong sunlight penetrates only where a large tree has crashed down, bringing down smaller trees on the way. In the "light gap" thus created, young trees spurt up, eventually to close the canopy once more. Most trees do not branch out near the ground; their leaves are best deployed higher up, where they have a chance of catching some light. The biggest problems are not impenetrable walls of vegetation, but fallen logs and slippery tracks. Rare is the walker who travels in completely uncharted territory. Most forests are crisscrossed with the tracks, however obscure, of earlier explorers, hunters, prospectors, rubber tappers, woodcutters, and so forth—not to mention the indigenous people of the forest.

Where a pristine forest has been cleared and left to regenerate, the new growth, called secondary forest, is bushier and harder to move through. Swamp forests are also hard to cross, for the obvious reason that the ground is waterlogged and often perilously unstable. In any forest reaching out to a handy tree for support can be dangerous, not only because of the animals that may be disturbed. Many trees protect themselves with vicious spines on leaves, branches, or trunks. The sandbox tree, so called because its fruit was popular for making boxes to hold blotting sand, has nasty spikes growing out of its trunk, sap that can blind a human, and fruit that explodes and sends its poisonous seeds flying for sixty feet.

The mass of life that is a tropical rainforest may seem at first a crowded jumble. But one can find an underlying order by studying its separate parts and their interconnections. The forest appears a confused tangle partly because of the climbers that cross from tree to tree and destroy any sense of regularity that might otherwise arise from the collection of straight, unbranched trunks. The epiphytes—plants that grow on tree trunks, on branches, on lianas (woody climbers), and even on leaves—add to the feeling of density. The forest's fullness is accentuated by its different "layers." Besides the ground plants and shrubs, the vines, lianas, and epiphytes, the trees themselves range in height from 50 to 160 feet, the smaller ones obscuring the giants from ground view.

The complexity of a tropical rainforest sometimes strains the crude framework of modern biology, derived from the much simpler ecosystems of the temperate zones. In order to begin to make sense of the richness of rainforest life, scientists have, somewhat artificially, resolved this continuum of trees into a number of strata, usually three or four. Each layer provides a distinct ecological niche, so that we see, in the phrase of Alexander von Humboldt, one of the greatest Amazonian explorer-naturalists, "a forest above a forest."

Visitors new to rainforests, primed by the purple prose that characterizes almost all writing on the subject, are often disappointed not to find the trees dripping with bizarre and brilliant flowers and the air reverberating with the cries and howls of exotic creatures. The forest does not advertise itself; the overwhelming impression is of a green stillness. Even Henry Bates, traveling with Alfred Wallace at the beginning of his eleven-year sojourn in Amazonia, felt let down. Writing of his and Wallace's first visit to the forest in 1848, he asks:

> But where were the flowers? To our great disappointment we saw none, or only such as were insignificant in appearance. Orchids are very rare in the dense forests of the lowlands. I believe it is now tolerably well ascertained that the majority of forest trees in equatorial Brazil have small and inconspicuous flowers. . . . We were disappointed also in not meeting with any of the larger animals in the forest. There was no tumultuous movement or sound of life. We did not see or hear monkeys, and no tapir or jaguar crossed our path. Birds, also, appeared to be exceedingly scarce.

Luckily for Bates, whose patrons had promised him and Wallace fourpence for each insect they collected, the forest was more heavily populated than it first appeared. "I afterwards saw reason to modify my opinion, founded on these first impressions, with regard to the amount and variety of animal life in this and other parts of the Amazonian forests. There is, in fact, a great variety of mammals, birds, and reptiles, but they are widely scattered, and all excessively shy of man." Many animals are strictly nocturnal, or live only in the trees. It is by its sounds that the life of the forest makes itself known to the outsider. The disquieting roars of howler monkeys fade out, to be replaced in turns by cries of unseen birds, the crash of a falling branch, the rustling of leaves as a small creature scuppers away, and other, unidentifiable, noises. The newcomer sees little movement or color. It helps to be quiet, to have an eye for detail, and to know where to look; it is even better to have a companion who knows the forest well.

. . .

The main canopy of the forest is formed by the crowns of the middle strata of trees, generally those from 100 to 130 feet high. A few taller trees of the forest emerge from this canopy, too widely separated to form a continuous layer. Below the canopy are the shorter trees, 50 to 80 feet high, and the relatively sparse understory, consisting of shrubs, nonwoody plants, seedlings, and young trees. The forest floor is often bare, save for a thin litter of dead leaves. Superimposed on this framework are climbers and epiphytes, plants that occupy all levels and even move back and forth from one to another.

The tallest trees in the world are not in tropical rainforests. The giant redwoods of California and the eucalyptuses of Australia, some of which are more than three hundred feet high, outstrip the emergents of the rainforest, which rarely reach two hundred feet. Neither are rainforest trees known for their great girth. The vast majority have relatively slender trunks. Larger trees are dotted about the forest —the biggest circumference recorded is fifty-six feet—but most are no more than three feet around.

Tree roots are more prominent in rainforests than in temperate forests. Many roots break through the surface and snake along the forest floor, posing a hazard to walkers. Most of the roots are shallow, and their inadequacy as anchors is brought home to anyone walking or sleeping in the forest, who, perhaps several times in a day or night, will hear the cracking of branches and the muffled thud that signals a tree hitting the ground. Because they are tall, slender, and shallowly rooted, many trees have evolved auxiliary means of support. Some have buttresses, giant flanges that flare out from the trunk and act as props. Frequently buttresses are so big—rising as high as fifteen feet and increasing a tree's girth by many times—that they protect a tree from human interference. Loggers cannot cut through the trunk at ground level; they are forced to work on a platform built overhead or abandon the job. On the other hand, the flatness and size of buttresses make them attractive to small-scale woodcutters, who have mutilated many in the process of cutting usable planks from them.

Perhaps the most curious root formations are the so-called stilt roots. These are roots that grow down from the trunk or branches of a tree. Palms often have stilt roots, as do many trees growing in swampy areas. Some stilts are straight, others arch out from the trunk in graceful bows, twenty or thirty surrounding the lower part of the tree and hiding it from view. The web of climbers that weave back and forth through a rainforest also serves to give some support to trees, although it has the opposite effect when a liana-entangled tree falls and pulls several others down with it.

Many rainforest plants can tolerate low sunlight, especially when they are young, and some even prefer deep shade. But most trees deploy their limited resources to ensure that they get as much sun as they can. Trees also have ways of conserving their nutrients so that they are not lost to animals or soaked away by rain. They use this store of energy efficiently. Entirely different species share certain characteristics useful in achieving these ends. The lowest branches of most canopy trees, for example, are seventy or more feet above ground, where there is enough sunlight to pay back, through a higher rate of photosynthesis, the energy required to form branches. Their leaves are often thick enough to store nutrients from rainwater, and they are tough and glossy in order to reduce leaching and transpiration (evaporation from leaves). Many tropical rainforest trees have similarly shaped leaves: oblong with a pointed "drip tip" that leads the rain off so that a pool of water does not collect on the leaf and leach away its precious store of nutrients. By keeping the leaf relatively dry, the drip tip also protects it from being colonized by moisture-loving lichens, algae, and mosses. Emergent trees, exposed continuously to the drying effects of full sunlight, do not need protection against damp and hence seldom have drip tips; but they are more likely to lose their nutrients to rain and transpiration, so their leaves tend to be even thicker than those of canopy trees. Most of the low-growing plants, living in deep shade and a very humid atmosphere, have large thin leaves that intercept more of the available sunlight.

Epiphytes and lianas depend on the forest's framework of trees and shrubs for physical support. By exploiting difficult niches, they contribute to the density and richness of the forest. Epiphytes are not usually parasites. They grow on the trunks, branches, and leaves of other plants, using them for support and collecting water and nutrients from their surfaces. But they do not harm their hosts. In fact, recent research has shown that many trees send out feeder roots from their branches and trunks into the organic matter that accumulates on the epiphytes they support. The biggest problem for epiphytes, with no access to the ground litter, is getting enough water and nutrients. They survive instead by extracting nutrients from the dead leaves and twigs that fall on them from the canopy. Some epiphytes require strong sun and keep to high branches; others grow in the depths of the forest, where it is shady and damp.

Orchids are among the most common epiphytes in tropical rainforests. They, like other epiphytes, have special structures that enable them to absorb water quickly when it is available, and that protect them from dehydrating. Many have special bulging stems that look like bulbs and hold extra water. They also have two kinds of roots. One sort anchors the orchid to its support; the other, long and free-hanging, is wrapped in a white, water-absorbing tissue that helps it scavenge water and nutrients from the air, the ground, tree trunks, and branches. Once their roots have absorbed water, the orchids store it in thick, waxy leaves that are less vulnerable to transpiration.

Bromeliads, members of the pineapple family, also have thick, waxy leaves. They are arranged in a tight rosette that forms a tank and catches rain and debris falling from the forest canopy. Eventually the plant develops its own compost heap, which provides it and the animals that nest in it with water and nourishment. Some of these "tanks" can hold more than twelve gallons of water. Other epiphytes, including ferns, have different ways, such as cup-shaped leaves, of catching water and debris. The miniature gardens thus formed support many typically terrestrial species, including cockroaches, earthworms, and snakes, high up in the forest canopy. And thanks to the pools of water in some epiphytic plants, even some aquatic species, such as salamanders and frogs, can live high in the trees.

Stranglers are bizarre plants that start out as epiphytes. They germinate in treetops and eventually put out long roots that anchor in the ground and send nutrients back up to the plant. With this boost, a strangler, most of which are in the fig family, expands rapidly. Its own crown starts to shade out the crown of its host tree, taking most of the available sunlight and water. The strangler's roots surround the host's trunk, smothering it and outcompeting its roots for nutrients. Finally, the support tree dies and decays, leaving a healthy strangler standing upright with a hollow center.

Climbing plants, called lianas or vines, are among the most striking features of rainforests. Many plant families, notably the passionflower, legume, and cucumber families, include climbers; even a few ferns are lianas. Most rainforest lianas are woody; their stems are variously shaped—some are circular, some twisted, some wavy and flat like ribbon, and some so contorted as to be almost curly. Lianas may grow to a great size, as thick as a man's body and hundreds of feet long. Several rattans, which are the longest lianas, grow to a length of more than five hundred feet. Certain vines, such as philodendron and *Monstera deliciosa* (the Swiss cheese plant), both common houseplants, and the deadly *Strychnos toxifera,* the source of strychnine, are partly epiphytic. In addition to their ground roots, they have smaller, sucker roots that attach to tree trunks for support.

Lianas begin life on the forest floor, and grow upward, using other plants for support, until they reach the upper level of the forest, where they finally branch out and flower, often spectacularly. They hold on to their supports by tendrils, hooks, and sucker roots, or they twine around them, though some rise straight to the top of the canopy without any apparent support. Climbers whose grip is too binding can kill or deform trees, especially immature ones.

Once at the canopy, they keep growing. Eventually their own weight pulls them down, but they continue growing, moving up, down, and along the ground, and forming labyrinths of loops and tangles with trees and other lianas. In the darker understory, climbers conserve resources by producing small leaves, keeping photosynthesis and transpiration to a minimum. But on reaching the canopy, some—*Monstera gigantea*, for example—develop much larger and differently shaped leaves and stems to take advantage of the sun. In full sunlight the leaves of *M. gigantea* grow to be six feet long, assuming a deeply cut, fernlike shape.

. . .

Few of the animals of a tropical rainforest move regularly between the floor and the canopy. Each species of mammal, bird, and insect lives mainly in a particular layer of the forest. Each layer—the floor, the understory, the canopy, and the giant emergent trees—is a distinct habitat, each with different conditions. So closely adapted are animals to their niches that in some areas, ornithologists estimate, as many as 95 percent of birds used to the darkness of deep forest will not cross a clearing even to breed or find food. The various levels of the forest are not easily distinguished from the ground, nor are the divisions clear-cut, but the differences are real and important. The upper stories get more rain and more sun. Full sunlight hits the emergents, but the forest floor, to which only 1 or 2 percent of the light penetrates, lives in dappled shade. In one Ivory Coast forest the maximum daily temperature in February is 82°F. at ground level and

90°F. in the canopy, 150 feet above ground. Daytime humidity on the forest floor is around 90 percent; at the top it is only about 60 percent.

More animals live in the canopy than in any other part of the forest. Half of the mammals (not including bats) in the rainforests of the Malay Peninsula, for example, live in the trees, compared with an average of 15 percent of mammals in temperate forests. Most canopy animals are herbivorous, but a few birds of prey inhabit the treetops; the spectacular harpy eagle of South America and the monkey-eating eagle (recently renamed the Philippines eagle by President Marcos, at the urging of conservationists who felt that the old name gave the bird a bad image). Their relative freedom from predators makes it safe for such canopy birds as toucans, hornbills, parakeets, and birds of paradise to be colorful and showy.

Many animals, insects especially, have shapes and colors that may seem flamboyant when seen out of context but that are effective camouflage in their own habitat. *Hymenopus coronatus,* for example, a Malaysian praying mantis, is shaped and colored like a tubular pink flower. In the canopy, on the end of a flowering branch, it is perfectly hidden from predators, but on the ground or in the understory, where bright flowers are rare, it would be highly visible and vulnerable. Likewise, the South American leaf-toad *Bubo pythonius* appears to be a dead leaf on the forest floor.

Tree-dwelling mammals—monkeys such as gibbons, howler monkeys, spider monkeys, and chimpanzees, as well as sloths, marsupials, squirrels, rats, and mice—rarely come to earth. They have strong claws, prehensile tails, or more specialized devices, such as the tarsier's adhesive pads, that give them a secure grip on their perches and enable them to swing or leap from branch to branch. Many tree dwellers, including some snakes, frogs, squirrels, and lizards, have expandable membranes that allow them to glide through the trees.

The understory has its own peculiar characteristics—low light, high humidity, a relatively confined space—to which many unrelated species have developed similar responses. In the deep shade of the understory, for example, both elephants and bellbirds rely on loud calls to communicate with their fellows. Tapirs, wild pigs, and anteaters all have long snouts, capable of rooting insects and fungi from the earth. The animals of the lower layers can climb, if at all, only to a limited extent, so speed and good camouflage are useful in escaping predators and in surprising prey. Many large mammals of the understory—striped cats and spotted deer, for example—move quickly and blend in well with the ground vegetation. Insects, especially, are adept at disguise and mimicry. Some butterflies have transparent wings—the ultimate in camouflage.

The mosquitoes that carry malaria and yellow fever generally feed on monkeys and other animals in the canopy of the forest where they live. Only when the forest is felled do they descend to the ground—away from the birds and bats that are their natural predators—to prey on humans and understory animals. Some viruses, such as those responsible for dengue fever, also live high in the canopy unless they are forced by tree felling to go to lower levels.

Insects make their nests in every conceivable part of the forest: on the undersides of leaves, under bark, in litter that collects on epiphytic plants, and on other arboreal creatures. J. K. Wagge and Robin Best, zoologists working at Brazil's National Institute for Amazonian Research, found that the three-toed sloth supports (and camouflages) various species of moths, beetles, ticks, and mites who live in its fur; several hundred insects may live on a single sloth. The sloth itself is camouflaged by the algae that grow on its fur and give it a protective green tint. There are many instances of insects having mutually beneficial relationships with the plants they nest and feed on. *Cecropia* is a common colonizing tree of Latin American rainforest clearings. Florists in New York sell its dramatically large dried leaves as decorations. It is also the only plant known to produce glycogen, an animal starch. Ants of the genus *Azteca* hollow out nests in the trunk and branches of cecropia. The ants live there with colonies of small insects that excrete a sugary solution that the ants eat. The tree also supplies the ants with glycogen and proteins from special growths at the base of the leaf stalks. Producing food for these ants diverts scarce resources from the tree's main job of reproducing itself. But the cecropia's expenditure is worthwhile, because the ants protect it from its enemies. They attack anything that disturbs the tree and kill climbers that might strangle it.

Several species of acacia trees, known as ant acacias, have a highly developed relationship with certain ant species. In the case of *Acacia cornigera* and the ant *Pseudomyrmex ferruginea,* each species depends absolutely on its relationship with the other. Like cecropia, these acacias are early invaders of cleared areas and are vulnerable to attack by leaf-eating insects and to being shaded out by faster-growing vines and trees. The tree has swollen thorns in which the ants hollow out nests. Its leaf stalks produce a nectar that provides the ants with the carbohydrates they need; special bright orange growths at the tips of the leaflets—called Beltian bodies, after the man who first identified them—supply the ants with protein and fat. The tree maintains the food supply by producing new leaves throughout the year, even during long dry seasons. A drought must last more than five months before the acacia tree loses all its leaves.

The queen of an acacia ant colony finds an unoccupied acacia seedling and burrows into a green thorn, where she lays her eggs. She feeds the larvae with the Beltian bodies and nectar but eats only nectar herself. Soon the new generation of workers takes over the job of foraging for food. Within nine months of the queen's arrival, the workers are patrolling the tree, walking up and down the branches and leaves day and night. They attack—biting and stinging—any other insects they find, and they kill any plants that grow within a thirty-inch radius of their tree. *P. ferruginea* is a good guard for several reasons. It is fast and agile enough to catch almost any insect. It has an extremely painful sting that is so feared by other animals and humans that many other insects mimic the acacia ant to frighten off predators. To humans, at least, *P. ferruginea* has a worse sting than do other related ant species that do not depend upon acacia trees. The colony may eventually number thirty thousand individuals. As it grows, it occupies all the thorns on the tree, and may expand into neighboring ant acacias, laying odor trails to link the separate sub-colonies.

Young ant acacias that do not have an ant colony suffer severe damage from other insects. In one study in Mexico, fewer than 2 percent of the shoots of an occupied ant acacia had other insects on them, but 38 percent of the shoots of an unoccupied tree were infested. In fact ant acacias depend on "their" ants for survival. "The experimental work with unoccupied swollen-thorn acacias indicates that if the acacia-ants were abruptly exterminated, the acacia population would be drastically reduced to the point of extinction in nearly all locations," wrote Daniel Janzen, the ecologist who first described the relationship. There is a direct correlation between the amount of energy an ant acacia expends on its ants and the degree of protection it gets in return. The more aggressive the worker ants are, the more effective they are in ridding the acacia of insect pests and of rival vegetation. But ants need food to fuel their aggressiveness, and 99 percent of the colony's solid food consists of the Beltian bodies. So the more food a tree produces, the more fierce will be its ants and the better protected the tree will be. The ants are equally dependent on their host trees; there is no substitute in their diet for the Beltian bodies.

Janzen thinks that in ant acacias the ants have replaced certain chemical and physical defenses that other acacia trees still have. No one knows exactly how non-ant acacias avoid being killed by insects and overshadowed by other plants, but their leaves are noticeably more bitter than those of ant acacias, and it may be that that makes them less acceptable to leaf-eating insects. The close relationship between certain acacia and ant species has influenced the evolution of other species as well. There are a few insects that can feed on ant acacias even when the ants are patrolling. They have evolved defenses against the ants, and they thrive in the predator-free environment the ants create.

The insects of the floor play a crucial role in breaking down forest debris, so that it can be recycled to provide nutrients to the living plants. One of the most common insects in rainforests is the army ant, of which there are at least 240 species. Army ants are semi-nomadic. Some species travel in packs of as many as twenty million members, foraging across a twenty-foot-wide front. They take the lichens, mosses, small bits of bark, twigs, leaves, and other insects they collect back to their temporary nests, where they break them down into food for their queen. These bivouacs may be nothing more than the linked bodies of the worker ants, surrounding and protecting the queen and her brood. The broken-down material that feeds the queen also nourishes the roots of nearby plants.

As the swarm advances, some of the insects in its path manage to run or fly away, often into the clutches of a number of species that have evolved to take advantage of army ant "leftovers." These camp followers include birds, beetles, springtails, millipedes, and flies. Some follow the odor trails the ants leave on the ground; others perch or hover overhead at different levels, catching flying insects in a well-defined order. Some depend entirely upon food that army ants "deliver" to them.

Many termites also forage in groups, but, unlike army ants, they are not nomadic. Their nests are up to seven feet tall and composed of chewed-up forest litter. The workers of one Brazilian species, *Anoplotermes pacificus,* build underground nests from chewed-up forest litter. Certain plants send roots into and around these nutrient-rich nests. The termites eat these root tips, an act that, although no one understands how, appears to benefit the roots, since they die if the termites abandon the nest.

Leaf-cutter ants, such as the South American species *Atta cephalotes,* also help the forest recycle itself. In single file they walk up tree trunks and out along branches to find suitable leaves. Each ant chews off a piece of leaf (often bigger than its own body) and carries it in its jaws back to an underground nest. There the ants produce a special food fungus by clearing the leaves of unwanted fungal spores, chewing them, and then planting them together with the desired fungus *(Rhozites gongylophora).* The leaf-cutters' larvae feed exclusively on this fungus, and the adults need the enzymes it produces in order to digest their main food, which is sap from the leaves they have collected.

R. gongylophora exists only in leaf-cutter nests. The relationship between the two species is so strong that the fungus no longer spends energy producing sexual spores; it relies entirely on the leaf-cutters to propagate it. New studies have produced more evidence of the close connection between the insects and the fungus. There are some plants from which *A. cephalotes* refuses to take cuttings. Scientists had until recently assumed that those plants contained chemicals that are toxic to leaf-cutters. But it is now clear that the leaf-cutters will also reject plants that are toxic to the fungus. This discovery has potential for agriculture and medicine, since it has led scientists to a number of new plant compounds that are active against certain harmful fungi.

. . .

Certain species of Latin American orchid are pollinated by a few closely related species of brightly colored bee, called euglossine bees. Because the two are so closely connected, the flowers are known as euglossine orchids. They emit an odor that attracts male euglossine bees from as far as a mile away. The bees have special brushes on their forefeet, with which they gather the fragrance from the orchids. But the bees do not get food from euglossine orchids. So why do they frequent the plants, and what is the advantage to them of having developed the highly specialized brushes? The answer turns out to be that the fragrance male euglossine bees are collecting acts as a signal, attracting male euglossine bees to one another. The males then form brilliant swarms, which in turn attract females, and the mating can begin.

Meanwhile, the orchids, each species of which depends upon just one or a few bee species, have evolved complex ways of ensuring that the bees carry out their unintentional mission of fertilizing them. Some orchids have slippery surfaces or an intoxicating fragrance, causing the bees to tumble deep into the pollen-producing part of the flower; others are structured so that while collecting the fragrance it wants, a bee receives a deposit of pollen on exactly the part of its body that will contact the female part of the next orchid the bee visits. These mechanisms not only ensure that the bee will transfer pollen correctly between flowers of one species, they also—in cases where two species of orchids are pollinated by a single species of bee—make certain that pollen from one species is not wasted by being bestowed on the female parts of another species. The need to attract euglossine bees, to load them correctly with pollen, and to minimize wasteful mispollination explains many of the bizarre shapes, colors, and scents that euglossine orchids have evolved and that have made them so fascinating to collectors.

Oropendolas are black birds with yellow tails, widespread in the forests of Latin America. Their nests—long, hanging, basketlike structures—are perhaps the most distinctive of any tropical bird. The cowbird, a sort of tropical cuckoo, lays its eggs in oropendola nests. Neal Smith, a zoologist working for the Smithsonian Institution in Panama, noticed that some colonies of oropendolas eject cowbird eggs from their nests and others do not. He wondered why this should be. Eventually he noticed that colonies that nest in trees that also have wasp or bee nests reject cowbird eggs, and that colonies that have no wasp or bee nests in their trees accept cowbird eggs. Still later, Smith found out that oropendola chicks are often infested by botflies. The botfly maggots eat the young oropendolas and are a major cause of death in oropendola chicks.

Smith began making connections: He found that bees and wasps keep botflies away from their trees. So the oropendolas nesting in trees without bees or wasps—those that accept cowbird eggs—suffer from botflies. Looking more closely at these infested trees, Smith discovered that young oropendolas sharing their nests with cowbirds are nine times less likely to have botflies than those growing up without cowbirds. Cowbirds hatch and open their eyes earlier than their oropendola nestmates. They are aggressive, well able to defend themselves, and the oropendolas, against botflies. So cowbirds are useful parasites if an oropendola lives in a tree without any bee or wasp nests. Oropendola chicks in trees without bees or wasps are three times more likely to survive if they nest with a cowbird. The interconnections do not end with the tree, the bees and wasps, the botflies, the oropendolas, and the cowbirds. It is becoming apparent that mites also play a role that affects the behavior of both the cowbirds and the oropendolas.

Mimicry is another phenomenon that binds different species together. In one morning on Barro Colorado, a forested island in the Panama Canal that serves the Smithsonian Institution as a research station, I saw a katydid that resembled a delicately veined green leaf; a moth aping a decaying leaf, complete with frayed holes in its wings; clumps of dead twigs that turned out to be insects; and, moving slowly across a leaf, a fly that looked exactly like an ant. Without the help of my companion, entomologist Annette Aniello, I would have overlooked or mistaken all of them.

"Everything here looks like something else," she remarked as we walked along. I paused, waiting for her to shake off a wasp that had just perched on her notebook. The wasp flew to a nearby leaf and I walked on, but Aniello stood still, peering at it. Suddenly she ran off, telling me to stay. I diligently stood guard over the wasp until she came back with the net she had left behind and scooped up her prey. She took another look at it and said excitedly: "I thought this was a spider wasp, but a wasp wouldn't have perched; it would keep moving. Then when I got a closer look and saw the mouth parts and the size of them, I knew it was a fly and a predatory one. I've never seen this one before—maybe no one has, but I know what family it's in, the Asilidae, which is full of things that mimic other insects—orchid bees, other bees, wasps. This is some kind of robber fly. It's obviously got a nasty bite, and it's mimicking a certain kind of spider wasp that gives a fierce sting. That's called Mullerian mimicry, when one dangerous species takes advantage of the bad reputation of another species by looking like it. After all, even toxic insects are better off alive, and the most efficient way of advertising the fact that you are dangerous is to use signals that are already well known.

"This fly is one of the best mimics I've seen. It's an interesting case, because there are two other insects in this series, mimicking the same wasp. One is the assassin bug, which gives a painful bite. And there's a katydid that looks just like the rest of them, but it's harmless. That's a case of Batesian mimicry, when something innocuous protects itself by imitating something nasty."

When we got back to the laboratory, Annette popped the newly collected fly into the freezer—"it's the best way of killing insects down here"—and checked the collection of the island's insects that she and earlier entomologists have made. I saw the spider wasp and its look-alikes, the assassin bug and the katydid, but the mimicking fly was not there. "I'll have to send this off to the Smithsonian to find out if it's a new species. It's never been recorded from here, that's for sure, but maybe it's known from somewhere else. If not, we'll name it. That'll be one more for science."

4

Boundless Fertility

Between 40 and 50 percent of all types of living things—as many as five million species of plants, animals, and insects—live in tropical rainforests, though they cover less than 2 percent of the globe. These forests are the richest regions on earth in terms of biological diversity and in terms of pure bulk, the mass of living organisms packed into a given space. The average biomass—that is, the dry weight of the plants and animals—of a tropical rainforest is 180 tons per acre; in temperate forests it is only 120 tons per acre.

Rainforests are distinguished not only by the sheer quantity of life that they support but by the diversity of that life, the number of different species of plants and animals. Tropical forests contain from 20 to 86 species of trees per acre, whereas a temperate forest has only about 4 tree species per acre. The forests of the North American temperate zone have fewer than 400 species of trees, whereas the Indian Ocean island of Madagascar has 2,000. For nonwoody plants the contrast is even greater. Great Britain, twenty times as big as the Pacific island of New Caledonia, has 1,430 species of flowering plants, compared with the latter's 3,000, 80 percent of which grow only in New Caledonia. One square mile of rainforest near the Colombian city of Quibdó has 1,100 species of plants. Tiny Panama has as many plant species as the whole continent of Europe.

Temperate countries are poor by comparison. Great Britain, for example, is twice the size of the Malay Peninsula. Yet it has only 1,430 plant species to the peninsula's 7,900. An area of Southeast Asia one quarter the size of Western Europe has 297 species of land mammals, more than twice as many as does Europe. In a three-hundred-square-mile area of forest on the border of Panama and Costa Rica, Edward O. Wilson and the late Robert MacArthur found more than 500 resident bird species, four times as many as there are in all the broad-leaved temperate forests of eastern North America. Mount Makiliang, a forested volcano in the Philippines, has more woody plant species than all of the United States. Ghillean Prance, a tropical botanist at the New York Botanical Garden, found 235 species of trees, counting only those whose trunk was at least two inches in diameter, on 2.5 acres in the Amazon forest. Daniel Janzen and Thomas Schoener captured 500 species of insects in two thousand sweeps of a net in the understory of a Central American rainforest, where insects are less numerous than they are in the middle and canopy layers.

A typical four-square-mile patch of rainforest, according to a report by the U.S. National Academy of Sciences, contains up to 1,500 species of flowering plants, as many as 750 species of tree, 125 species of mammal, 400 species of bird, 100 of reptile, 60 of amphibian, and 150 of butterfly, though some sites have more. Insects in tropical rainforests are so abundant and so little known that it is difficult to establish an average density. The same report cites a recent estimate that 2.5 acres might contain 42,000 species. Ten square feet of leaf litter, when analyzed, turned up 50 species of ant alone. Terry Erwin of the Smithsonian Institution conducted detailed surveys of the insects on tropical trees. He estimated that each species of rainforest tree supports 405 unique insect species. Erwin's figures indicate that there may be 20 million or more species of insect alone in tropical rainforests, many times more than previously suspected. More recent work by Sir Richard Southwood, head of Oxford University's Department of Zoology, suggests that Erwin's estimates may be too low.

A high proportion of animals and plants in tropical rainforests are endemic to one area—that is, they live nowhere else. This is especially true of Southeast Asia and Oceania, with its over twenty thousand islands. Papua New Guinea has 320 endemic species of birds, almost half its total. There are 180 mammal species in the

Philippines, and over half are endemic to that country. Sixteen percent of all the species of birds in the world occur in Indonesia (1,480 out of 9,000), and almost one quarter of them are endemic. Forty percent of all birds of prey depend on tropical rainforests. According to the conservation monitoring unit of the International Union for the Conservation of Nature (IUCN), an association of governments, scientific associations, and conservation organizations, the majority of all insect, reptile, and amphibian species occur in tropical rainforests.

Rainforest species have what one authority calls "a tendency to gigantism." Richard Spruce, the sickly Yorkshire schoolteacher whose fifteen years in Amazonia made him one of the great Victorian founders of tropical botany, remarked in a letter home that "nearly every natural order of plants has here *trees* among its representatives. Here are grasses (bamboos) of forty, sixty, or more feet in height. Vervains forming spreading trees with digitate leaves like the Horse-chestnut. . . . Instead of your Periwinkles we have here handsome trees exuding a milk which is sometimes salutiferous, sometimes a deadly poison, and bearing fruits of corresponding qualities. Violets of the size of apple trees."

Animals show the same tendency, as the names of the African giant snail, the bird-eating spider, and the goliath beetle indicate. The electric blue Morpho butterflies, which are among the great joys of the South American rainforests (the naturalist Frank Chapman described them as "the bluest things in the world"), have wingspans of up to seven inches. The Queen Alexandra birdwing butterfly and the owlet moth both have twelve-inch wingspans, bigger than that of many birds. There are eleven-inch-long millipedes in the Seychelles Islands, and slugs of the genus *Vaginulus* that are eight inches long.

With so many different species, there can be relatively few individuals of each. So, compared with the plants and animals of temperate forests, most tropical forest species are rare. Most plant species occur less than once in an acre. A survey of five acres of forest in the Brazilian Amazon found 1,986 plants over five feet high, representing 502 species. This means there was an average of only four individuals of any one species. A sixty-acre patch of rainforest on the Malay Peninsula has 381 species of tree, of which 157 occur only once. In the words of Alfred Wallace:

> If the traveller notices a particular species and wishes to find more like it, he may often turn his eyes in vain in every direction. Trees of varied forms, dimensions and colours are around him, but he rarely sees any one of them repeated. Time after time he goes towards a tree which looks like the one he seeks, but a closer examination proves it to be distinct. He may at length, perhaps, meet with a second specimen half a mile off or may fail altogether, till on another occasion he stumbles on one by accident.

Many rainforest species have a very limited distribution; some birds, for example, live only on one island or one mountain range. This rarity itself makes rainforest species particularly vulnerable to extinction. The combination of rare species and threatened habitats means that some species have disappeared before they were ever known to man. There are probably ten thousand to fifteen thousand species of flowering plants in South America alone that are still unknown to science. And South American rivers have perhaps two thousand as yet unnamed species of fish. In all likelihood many of these will become extinct before they are found by science.

One reason for the diversity of life in rainforests is their great age. Evolution has rolled on in many rainforests for the past sixty million years, making them the oldest communities on earth. The richest rainforests are in Southeast Asia, where the climate has remained stable and favorable to life—tropical and moist—for many tens of millions of years. There animals such as the wormlike *Peripatus* (a link between two major groups of invertebrates—worms and caterpillars) and such plants as the gingko tree have survived almost unchanged since ancient times, living fossils. West Africa, by contrast, has endured long dry periods during the Pleistocene epoch and is floristically the poorest of the three main regions. On the other hand, creation of new niches by periodic upheavals, as mountain ranges and rivers formed and disappeared, earthquakes struck, and volcanoes exploded, also encouraged the evolution of new species. When the population of a species is divided and then isolated—by the disappearance of a land bridge between islands, by the creation of a mountain range, or by the fragmentation of forests—natural selection takes different courses in each population and new species may evolve. Southeast Asia owes its species richness, therefore, both to its stability and to the creative disruption of its landscape.

Another reason for the extraordinary richness and diversity of life in tropical rainforests is the hot and moist climate. Because temperatures never drop to the freezing point, organisms can grow and reproduce continuously throughout the year. Survival depends not upon enduring periods of extreme cold or drought but upon finding an ecological niche in which one can hold one's own. Being able to adapt to changing circumstances as competing species evolve is all-important. A continuous growing—and breeding—season speeds up natural selection, so species can evolve more quickly in response to changes in their environments. This effect is greatest among invertebrates and cold-blooded vertebrates; an individual of the same species develops much more quickly in the warm rainforest than in temperate zones. In North America the butterfly *Danus chrysippus* completes its life cycle in a year, but in the Philippines it takes only twenty-three days. The beetle *Crioceris subpolita,* a Javanese species, matures in one month; its German cousin, *C. asparagi,* takes one year to complete the same process.

The climate provides an abundance of ecological niches. The transition from aquatic to terrestrial life probably occurred in just such warm, moist conditions. Rainforests are the only places on earth where typically aquatic animals can live out of the water; terrestrial leeches are probably the best-known examples of this phenomenon. Even marine species—fishes such as the mud skipper and climbing perch, and certain worms and crabs—can live on land in rainforests.

By specializing, plants and animals avoid competition. Specialization also means that the forest's resources are very efficiently exploited, which enables it to support more species and individuals. The wild relatives of many important crops—including rubber, coca, pineapples, cashews, and mangoes—can only be pollinated by certain species. One of the most fantastic cases of specialization and coevolution between plants and pollinators involves a passionflower *(Passiflora kermesina)* and the butterfly *(Heliconius ethilla)* that feeds on the nectar of the passionflower and fertilizes it.

The butterfly lays its eggs on the vine, but the caterpillars eat it, so the plant has evolved various ways of discouraging its pollinator from laying eggs on it. The passionflower manufactures toxic chemicals to make itself inedible. But the butterfly has developed the ability to absorb the chemicals; in fact, they protect it from predators. The vine has another defense—small, edible, egglike growths on its tendrils that confuse the larvae into eating their own eggs. This ruse doesn't always work, however, because the butterfly has learned to distinguish false eggs from real ones. The plant also has hooked leaf hairs that kill butterfly larvae by puncturing the abdominal cuticle. But the larvae have evolved special pads that prevent such damage. Still evolving, the vine has developed false tendrils on which the butterfly may mistakenly lay its eggs but which are so weak that the eggs drop off before hatching. Near the spots where butterflies are most likely to lay their eggs, the plant secretes a nectar that attracts ants and other egg predators. The butterfly responds by being able to detect vines that are already infested with predators, and by laying eggs on the tips of the tendrils, away from the ants.

One disadvantage of highly specialized species is that, because they depend on a particular set of conditions, they are more likely to become extinct if their environment is disturbed. The United States imports more than $16 million worth of brazilnuts every year, gathered by Indians and peasant collectors from trees scattered throughout the forest. Some years ago an entrepreneur decided that it would be more efficient to grow the nuts on a plantation. The trees were planted, they grew well, and in due course they flowered. But they produced no nuts. No one knows how brazilnuts are pollinated, but they are visited regularly by euglossine bees, the same ones whose mating rituals depend on certain chemicals available only from a few species of epiphytic orchids. The plantation had none of these orchids, nor did it have the other plants on which the bees depend for food. No bees, no brazilnuts. Brazilnut trees must grow in a mixed forest, one big enough to encompass all the other species involved in the life cycle of the trees. Exactly what they are is still unclear, but they include not only the euglossine bees and the plants and insects they may need (such as those that pollinate the epiphytic orchids), but also the rabbit-size rodents called agoutis, which are the only creatures that crack open the hard fruit case that contains the valuable brazilnut. The seeds need agoutis to free them from their outer coverings. As yet no one knows what plants and animals the agoutis depend upon.

. . .

Tropical rainforests have adapted to very difficult environmental conditions. They are subject to constant high temperatures and frequent and intense rainfall. Subtract the forest from the ground on which it stands, and with few exceptions you are left with poor soil —often on steep slopes. These conditions have doomed most attempts to convert rainforests into profitable and enduring farms and plantations. Yet with the forest in place, this hostile environment supports the most productive ecosystem on earth.

Why is rainforest soil so poor? Partly because it is so old. Soil is a mixture of air, water, decomposed vegetation, and broken-down rock. Its fertility depends largely on the quality and age of the rock from which it has formed. The Amazon Basin developed between two ancient rock masses, the Brazilian and the Guayanan shields, several billion years old. They are among the oldest rock formations on earth, and the soil formed from them is ancient, weathered, and infertile. The Amazon rises in the Andes, which are only some twenty million years old, so the soil it deposits on its floodplain is young and fertile. But many other Amazonian rivers, including the Tapajos and the Trombetas, rise in the shields and carry infertile silt.

Almost two thirds of the humid tropics have soils that are acidic and low in nutrients. These soils, technically known as oxisols and ultisols, but formerly and still commonly called tropical red earths, are unsuited to orthodox temperate agriculture. This is because they are deficient in most nutrients and waste any artificial fertilizers that are added. They tend to immobilize phosphorus, which combines with the high levels of iron and aluminum in such soils to form extremely insoluble compounds that plants cannot use. Other nutrients, especially those that are positively charged—such as calcium, magnesium, and potassium—tend to pass right through tropical red earths as soon as they are dissolved by rain. Only a fraction are absorbed by plants before they are leached away. The most publicized feature of tropical red earths is their tendency to turn into laterite, a hard, bricklike clay, when they are cleared of trees and exposed to the sun. But in fact only a small proportion of tropical red earths react that way when cleared.

Scattered throughout such regions as Amazonia that are dominated by tropical red earths are small patches of alfisols, another soil type that is more fertile and less acidic. They cover only 4 percent of the humid tropics but are better for conventional agriculture.

The other group of fertile rainforest soils comprises those too young to have lost their mineral nutrients through thousands of years of leaching. Volcanic regions of the Philippines, Indonesia, Papua New Guinea, Central America, and Cameroon, as well as parts of the floodplains of the Amazon and the Congo River systems, have these more fertile soils. They account for around 18 percent of the soils of the humid tropics. For the most part, they already support dense human populations.

About 12 percent of the humid tropics have soils that, although young, are sandy, very acidic, and even less fertile than the tropical red earths, or they are very shallow and highly prone to erosion. These soils are the least suitable for orthodox agricultural development.

The humid tropics of America have the worst soils; 82 percent are acid and infertile, compared with 56 percent in the African humid tropics and 38 percent in Asia. Forty percent of humid tropical Asia has somewhat fertile soils, as does 24 percent of Africa and only 13 percent of Latin America. On the other hand, the most difficult soils, those that are very shallow or extremely acid, cover only 5 percent of the American tropics, whereas Africa is plagued with them on 20 percent of its land and Asia on 22 percent. In the Amazon Basin, which contains more than one third of the world's tropical rainforests, only 6 percent of the soils have no major limitations to agriculture. Six percent of Amazonia is still eighty million acres, but those eighty million fertile acres are scattered throughout the basin in, as it were, unmarked plots.

. . .

In temperate forests practically all the nutrients are in the soil. A study conducted in a rainforest at San Carlos de Río Negro in the Venezuelan Amazon found that three quarters of the nutrients were in the biomass—the living plants and trees themselves; 17 percent were in the forest debris (the humus and litter); and only 8 percent were in the mineral soil. The tropical rainforest is a closed system, within which the same nutrients are continually recycled. It is a system so efficient that there is hardly any loss or waste. As soon as a leaf or a branch dies and falls to the ground, it begins to decay. Microorganisms attack the debris and speed up the process of decay, and specialized roots help the plants absorb the nutrients as soon as they are released. Decay and uptake are so fast that the litter layer is seldom more than an inch or two thick.

Because a high proportion of their nutrients comes from above, rainforest trees have many small "feeding roots" that spread out on the forest floor and even grow up trunks to capture nutrients from water trickling down. Often in rainforests a thick, spongy mass of roots, fungi, humus, bacteria, and other microorganisms covers the soil. Here in the root mat, which may be as much as sixteen inches thick and can be peeled back like newly laid lawn turf, the forest decomposes and nourishes itself, acting as a slow-release fertilizer.

When a leaf or twig falls onto this mat, thin root hairs quickly invade it and absorb nutrients as the leaf decays and releases them. The roots of many plants are invaded by specialized fungi, called mycorrhizae, that absorb minerals and water more efficiently than uninfected roots. These "fungus roots" can extract such nutrients as phosphorus even when they are present in concentrations too low for ordinary roots to exploit. Scientists are just discovering that in tropical rainforests, where soil fertility is low, many, possibly most, trees cannot grow or survive without mycorrhizae, many of which are specific to particular species. When a rainforest is burnt or the soil otherwise damaged, so that its mycorrhizae are destroyed, the forest may never be able to return. A 1982 report by the U.S. National Academy of Sciences calls mycorrhizae "the cornerstone of mineral conservation by natural tropical forests" and says that they are "often more cost-effective in the long term" than artificial fertilizers. Ironically, fertilizer suppresses the growth of mycorrhizae, so each subsequent application must be increased to compensate for the loss of mycorrhizal nourishment. Foresters in Puerto Rico tried for twenty years to establish pine plantations. They experimented with twenty-six different species and varieties, but all their attempts were in vain, until they inoculated the trees with mycorrhizae from a natural pine stand. But at present we cannot cultivate these fungi, nor are we able to transfer them to uninfected plants on a large scale. We must rely on their natural presence in forest soils.

In San Carlos de Río Negro, Venezuelan scientists added radioactively marked calcium and phosphorus to the root mat of a tropical rainforest. They then collected the rainwater that drained through the mats and measured how much of the "tagged" material had leached through and been lost. Over 99 percent of the calcium and phosphorus had been absorbed by the root mat; less than 1 percent was leached out by the rain. In other words, the nutrient recycling system of an intact tropical rainforest is almost 100 percent efficient, losing nothing to the outside world.

A tropical cloudburst can drop one inch of rain in half an hour —up to forty times more water than the average shower in the northeastern United States. A typical storm in Ghana drops eight inches of rain in one hour, four times as much as London gets in an average month. The forest is its own protection from the devastating pressure of tropical storms. Many tropical forests grow on mountains and hillsides. Their canopies cushion the soil from the impact of the rain, protecting it from erosion and landslides. Their roots act as sponges, aborbing the rain and releasing it slowly. This way the forests to some extent even out seasonal extremes, conducting a steady and moderate flow of rainwater to the world's major rivers. When the forest is removed, so is this moderating influence. During the rainy season the full force of tropical storms is felt at once, and there are no reserves left to ease the hardship of the dry months. With cruel irony, deforestation brings flooding *and* drought.

From half to three quarters of the rain that falls on a tropical rainforest is intercepted by the forest canopy and eventually evaporates back to the atmosphere. The rest trickles down branches and trunks or drips from leaves. Leaves scavenge nutrients from rainwater; they support colonies of algae, lichen, and liverworts, which absorb minerals from the water that falls on them. The nutrient-rich rainwater also seeps through the characteristically thin bark of the rainforest tree to the living tissue beneath.

Erosion, flooding, and landslides resulting from forest clearance affect people downstream of the deforested area as much as they do those who are working the land. According to the World Wildlife Fund, 40 percent of farmers in the developing world live in villages that depend upon the rainforest sponge effect to prevent or mitigate flooding and drought. A single storm in the humid tropics can sweep

70 tons of soil from one acre. The Mississippi River carries only 265 tons of soil per square mile, whereas each square mile of the Ganges is loaded with 4,200 tons of soil washed down from the Himalayas. In the Ivory Coast one acre of forest on a 7 percent slope loses less than a hundredth of a ton of soil a year. But once the trees are removed, the same amount of rain falling on the same area washes away 60 tons of soil.

Erosion on that scale has shortened the life of many dams in the tropics. According to the chairman of India's National Committee on Environmental Planning, "the average rate of sedimentation in most reservoirs is 4 to 6 times as high as the rate which was assumed at the time they were designed and built." Water shortages caused by premature sedimentation of reservoirs, and a 250 percent increase in typhoon damage in the past twenty years, made Philippine President Marcos declare the destruction of the Philippines forests "a national emergency."

In the Himalayas the cost of deforestation is reflected in the way the monsoon storms are, year by year, killing more people, destroying more property, and costing individuals and the state more money. Twelve percent of the human race—500 million people in India, Pakistan, and Bangladesh—depend on water from the Himalayan watershed. The deciduous rainforests of the Himalayas once regulated the water flow and gave natural protection against the annual storms, but those forests have been severely depleted by logging. Almost half the forests that covered the Himalayas in 1950 have been destroyed by commercial loggers. Between 1971 and 1979, 2.3 million people crossed illegally from Bangladesh to India to escape flooding and drought resulting from deforestation. India, whose own deforestation problems are just as severe, no longer accepts them, and Bangladesh refuses to have them back. According to India's National Commission on Floods, the area in India affected by annual floods grew from sixty million acres in 1950 to one hundred million acres in 1980. Six billion tons of soil are swept from the hills each year; the value of the lost nutrients alone (five million tons of nitrogen, potash, and phosphorus) is over $1 billion.

Landslides are common catastrophes: In Nepal twenty thousand were recorded in one day, wiping away laboriously maintained ancient rice terraces and burying whole villages. Rice yields have dropped by 20 percent in the last five years; maize yields by 33 percent. Down in the valley the runoff is raising riverbeds six to twelve inches a year. Flooding gets worse each year as the rains stream down the bare mountainsides into water channels too choked with eroded soil to contain them. The annual cost of damage from these floods in the mid-1970s was $1 billion. Deforestation has also worsened the desperation of the dry season and made it more difficult to improve the living standards of villagers. A $150-million pipeline intended to supply water to Nepalese villages lasted only eleven years before deforestation distorted the river flow so that there was not enough water in the dry season.

. . .

It is a widespread myth that rainforests produce a large proportion of the earth's oxygen, that they are the "green lungs" of the planet. In fact mature forests are in equilibrium. They consume as much oxygen in the decay of organic matter as they produce through photosynthesis.

The connections between deforestation, the world's carbon dioxide (CO_2) budget, and global climate, however, are more controversial and may be much more serious. Many scientists studying climate believe that burning tropical forests could change the global climate and destabilize the polar ice caps. If they are right, then destroying rainforests could ultimately cause Siberia to replace Kansas as the world's major grain-producing region and put London and New York under sixteen feet of water.

This hypothesis is based on the fact that rainforests are reservoirs of carbon, which is stored in the living vegetation. Releasing this carbon by cutting trees and then burning them or leaving them to decay will add to the concentration of CO_2 in the atmosphere. In the past century the amount of CO_2 in the atmosphere has increased by about 15 percent. Half the increase has occurred since 1958. The major cause is the burning of fossil fuels. Forest clearance could, according to George Woodwell, the noted marine biologist and director of the Ecosystems Center at Woods Hole, Massachusetts, account for as much as half of the added CO_2, though the figures are far from certain. Scientists agree that this steady increase will lead, probably in the next fifty years, to a doubling of the amount of CO_2 in the atmosphere.

Carbon dioxide and other elements in the atmosphere trap heat that would otherwise pass through the atmosphere into outer space. This "greenhouse effect" means that the predicted doubling in atmospheric concentrations of CO_2 will result in an average rise of about 5°F. in the earth's temperature. The warming will not be uniform around the globe; the North and South poles are forecast to have temperature increases of as much as 18°F., which may cause some melting of the polar ice packs, raising the ocean levels and changing rainfall patterns around the world. Such changes would alter present patterns of agriculture: The grain belt of the midwestern United States and southern Europe might become unsuitable for conventional temperate agriculture, whereas parts of India, Southeast Asia, and Siberia could benefit agriculturally from a warmer, wetter climate. The oceans could, as a result of the warming, release more CO_2 to the atmosphere, starting a feedback mechanism that would accelerate the disaster. Scientists by no means universally accept this view of the future. There are too many uncertainties and areas of ignorance to be sure what will happen as carbon dioxide builds up in the atmosphere. But, as Woodwell has pointed out, by the time we know what the effects are to be, it will be twenty years too late to do anything about it.

We already know that deforestation does significantly affect local climate. When large areas of rainforest are destroyed, less rain falls on the deforested region. Deforestation removes the sponge that holds rainwater and sends half of it back to the atmosphere directly by transpiration. Research in Brazil has shown that at least half of the rain in the Amazon Basin comes from water evaporated from the forest itself. In central Panama, where practically all the rainforests have been cut down to make way for cattle ranches, records show that annual rainfall has dropped by seventeen inches over the past fifty years. The cloud forests (rainforests that grow on mountains and intercept moisture from clouds) north of Veracruz, Mexico, collected water from coastal fog banks. When they were cleared, this source of moisture was lost to the region. Today more than one thousand square miles of once-rich land has been ruined; only desert vegetation will grow there. Some scientists have postulated that deforestation, which causes higher ground temperatures, will result locally in more intense updrafts and harder and more destructive, though less regular, rainfall. In addition, air heated up by increased ground temperatures would change wind and cloud patterns that could affect a much larger area than the deforested region itself. Deforestation in South America, for example, might alter the weather in North America. And a 50 percent drop in rainfall could render useless the elaborate system of hydroelectric dams that Brazil is planning for Amazonia.

. . .

So successful has the forest been in overcoming the limitations of its environment that it has invited its own destruction by people who have misread adaptation for boundless fertility. Even as great a naturalist as Henry Bates misjudged the Amazon, "a region almost as large as Europe, every inch of whose soil is of the most exuberant fertility." Theodore Roosevelt, on his collecting and big-game journey through South America, said of Amazonia, "Surely such a rich and fertile land cannot be permitted to remain idle, to lie as a tenantless wilderness, while there are such teeming swarms of human beings in the overcrowded, overpeopled countries of the Old World." Roosevelt acknowledged no serious obstacle to his vision of the future: "The very rapids and waterfalls which now make the navigation of the river so difficult and dangerous would drive electric trolleys up and down its length." Wallace was equally enthusiastic and unrealistic: "When I consider the excessively small amount of labour required in this country, to convert the virgin forest into green meadows and fertile plantations, I almost long to come over with half-a-dozen friends, disposed to work, and enjoy the country; and show the inhabitants how soon an earthly paradise might be created."

Yet long before Europeans formed their illusions about the Amazon, their ancestors in the Mediterranean were dismayed at the results of deforestation. More than two thousand years ago Plato wrote:

> There are mountains in Attica which can now keep nothing but bees, but which were clothed not so long ago with fine

> trees, producing timber suitable for roofing the largest buildings; the roofs hewn from this timber are still in existence. . . . The annual supply of rainfall was not then lost, as it is at present, through being allowed to flow over a denuded surface to the sea. It was received by the country in all its abundance, stored in impervious potter's earth, and so was able to discharge the drainage of the hills into the hollows in the form of springs or rivers with an abundant volume and wide distribution. The shrines that survive to the present day on the extinct water supplies are evidence for the correctness of my hypothesis.

Many people regard the warnings of ecologists about the consequences of destroying tropical rainforests as speculative and alarmist. Some see critics of the present state of affairs as trendy gloom-and-doom merchants overstating their case to justify an obsession for turning forests into untouchable museums of ecology. But such warnings have a long history. They have followed the axeman around the globe as he successively destroyed the forests of the Mediterranean, Europe, and North America. In fact, judging from the antiquity of the laments, one might expect an understanding of the mechanics of erosion and soil fertility to be part of the collective unconscious. "Often indeed," wrote Pliny the Elder, "devastating torrents unite when from hills has been cut away the wood that used to hold the rains and absorb them." In *De Re Rustica*, Columella, also writing in first-century Rome, explained that yields of crops planted on newly cleared forestland plummeted within a few years because the soil needed nourishment from the roots and leaves of trees.

. . .

When a tree dies and crashes to the ground, it pulls and knocks others down with it. A single tree may open an acre, and that space is enlarged as winds blow down the newly exposed trees. Suddenly a part of the forest floor is open to the sun, and the nutrient level in that small place gets a big boost from the new decomposing vegetation. Seedlings and saplings that have survived for years in deep shade spurt up. Seeds germinate as the soil in which they lay warms up. Within forty or fifty years a gap in the natural forest, one that has not been farmed or burned, has become difficult to distinguish from the surrounding forest in terms of bulk, although the species mix may be different from a mature forest.

Events will take a different course, however, if an area is repeatedly cleared for farming or burned before it is left to grow back. First of all, if the clearing is too large—more than 360 square yards, according to the FAO—the true primary forest species will not be able to recolonize it. Because there is little wind in the lower stories of the forest, most rainforest trees depend on animals and birds to pollinate them. Some pollinators are so adapted to the shaded, moist forest that they will not enter large clearings. The fruits of many rainforest trees are too large to be blown about by the wind; unless an animal moves them, they remain where they fall, near the parent tree. The vacant space will be invaded instead by weedy, colonizing species.

Burning and farming also destroy the root mat and the mycorrhizae and reduce the ability of seeds to germinate, thus damaging the forest's ability to recuperate. One study counted fifteen germinated seeds per square foot on a site that had been burned, and seventy per square foot in the undisturbed forest. Scientists estimate a farmed site will take well over one hundred years to return to the complex mix of species characteristic of the primary forest. If the site has been severely damaged, it may never recover. But no one knows for certain how long it takes a forest to reestablish itself. Large areas of forest around Angkor Wat in Cambodia that were cleared six hundred years ago and left undisturbed ever since are still distinguishable from the surrounding primary forest.

In 1971 a fifteen-acre section of forest in Venezuela was bulldozed for a military post. The post was never built, but the site is still bare today. According to Christopher Uhl, who studied the site as part of a United Nations–funded research project, "It may require millennia to revert to forest. The absence of any vegetation in this area appears to be a result of the three or four weedings to which it was subjected after being bulldozed. These weedings completely exhausted the seed bank in the soil."

Most tropical rainforests are felled so that the land can be farmed. Quite often the trees are left lying where they fall, or they are burned because it is impossible or too expensive to get them to potential buyers. Burning the vegetation has one good side effect: It releases the nutrients that are stored in the trees and gives a big (though short-lived) boost to the first crops planted. But when the trees are burned, the soil under them is also burned, or at least heated. When soil temperature exceeds 75°F., humus, the organic part of the soil, is destroyed faster than it can form. At higher temperatures those nutrients, particularly nitrogen and sulfur, volatilize and dissipate in the air. This happens when the soil loses its protective canopy and is exposed to the sun, but the damage is even greater when the forest is set afire.

In addition, once the trees are felled and the root mat destroyed, the soil is subject to the full force of the driving tropical rains. Heavy rain, first of all, removes nutrients by washing away the thin top layer of soil and leaf litter and by leaching nutrients deep into the subsoil, where they are unavailable to plant roots. It also compacts the soil, squeezing out the air pockets. Air is as important to soil quality as mineral nutrients, and compacted soil can be difficult to rehabilitate.

With the initial injection of rich ash from the burned vegetation, crops planted on newly cleared rainforest land may thrive for a few seasons, but productivity declines rapidly as the nutrient store is depleted. In Honduras, which has less rain than many parts of the humid tropics, yields of maize average nine hundred pounds in the year that a rainforest is cleared. But two years later, the average yield is only half that. Some rainforest soils, such as the porous sands of the Wallaba Forest in Guyana, are so poor that it is impossible to obtain even one harvest of an annual crop.

Most predator species in tropical rainforests are spread thinly over a large area, as are their food supplies (the particular plants or animals on which each depends). For this reason numbers never reach the critical level that may lead to epidemic damage. But the safeguard of low species density does not exist on farms cleared from rainforests where pest buildup is the second main cause of low productivity. Using pesticides on rainforest land may actually be counterproductive, because the relatively fast pace of evolution in the tropics increases the chances of immune "super-pests" evolving, as has happened in the case of malarial mosquitoes and DDT.

In temperate climates farmers routinely apply artificial fertilizers to their soils. But this is not usually practical on rainforest soils for several reasons. First, such fertilizers are expensive and in short supply in developing countries—especially in the remote regions where rainforests still exist. Second, the most common soils of rainforests, the tropical red earths, are unable to use artificial fertilizers efficiently because of their tendency to "lock up" phosphorus and to leach out such positively charged nutrients as calcium, magnesium, and potassium. Third, fertilizers may backfire by destroying nature's own growth promoters, the mycorrhizal fungi, and by boosting weeds that compete with crop plants.

. . .

What scientists know about tropical rainforests serves above all to convince them that we are still deeply ignorant about them. In the words of Seymour Sohmer, a tropical biologist and curator of the herbarium of Honolulu's Bishop Museum, "We should never lose sight of the fact that we know little or nothing about the way most humid tropical forests are structured and how they function, not to mention the component species and how they evolve." Most of the laws governing the behavior of tropical forests—when trees flower and fruit, how much light, water, and warmth they need to germinate and grow, how long they live, how they are pollinated, how a cleared forest regrows, how big an area each animal species needs to support itself—all these are still mysteries. It has been said that we know more about some areas of the moon than we do about tropical rainforests. Yet what we have learned about them so far has revolutionized our view of all life on this planet. The theory of evolution through natural selection, for example, is arguably the most important concept in the science of biology. This crucial concept arose from studies made by Alfred Wallace and Charles Darwin not in the biologically impoverished temperate zones but in the vastly more diverse tropics.

Scientists who set out to learn how rainforests work face greater practical problems than do those working in less complex ecosystems. Someone studying grassland could take as a representative area a patch several yards square and a few inches high. To get a fair sample of rainforest life one would have to observe at least ten acres to a height of 130 feet or more. So much of a forest's life takes place in the upper levels that someone on the ground misses most of it. Many tropical biologists today have to teach themselves to climb trees 160 feet high. Some live for months on tiny platforms 100 feet above ground in the forest canopy, moving gingerly between trees on jerry-built rope walkways or gliding with pulleys along rope webs.

Even climbing a tree in a tropical forest presents problems that don't exist in temperate regions. The lowest limbs of many trees are sixty to one hundred feet above the ground. The trunks and branches are full of animals—including snakes and innumerable insects, many of which are poisonous, aggressive, and hard to see. The wood of many species is so hard that metal spikes cannot pierce it. Donald Perry, a biologist working in Costa Rica, in 1974 developed a way of scaling rainforest trees that is now widely used in the tropics. "The method I settled on was to use a crossbow to shoot a fishing line over the tree's crown, and then to raise a strong rope into the canopy. I climbed the rope directly, using mountain climbers' devices called ascenders—special clamps that slide easily up a rope, then lock into place so that they cannot slip back down. One ascender supported a harness while the other supported two stirrups for my feet. Standing, I could raise the harness ascender. Then, sitting in the harness, I could pull up the lower ascender, moving my feet further up the rope."

In Brazil I tried a variation of this, using a different ascender for each foot, with the harness linked loosely to one of them. Rob Bierregaard, a rotund but evidently strong-armed ornithologist working for the World Wildlife Fund, seemed to have no problem jockeying himself up into the canopy and gliding gracefully down. With me, alas, it was different. Each minuscule advance was accompanied by a great deal of flailing about, grunting, and embarrassed swearing. In order to produce enough slack to allow the ascender to be edged upward, one must first raise the foot connected to the ascender. Dangling in the air, with nothing to push off against and plenty of time to look back down, the folly of attempting to break the laws of nature in this way becomes apparent. Once I reached my summit—which was considerably lower than everybody else's—I looked down from the branch on which I was sitting and contemplated the courses of action open to me. Eventually, deciding that a gallant death was to be preferred to public humiliation, I jumped off my perch. The descent, freefall, kicking off trunks or branches that came too close, was exhilarating.

My partner in this lesson was a German botanist who wanted to be able to climb so that she could discover how brazilnut flowers are pollinated. Once she learned the ropes, so to speak, she would be able to pull herself, and her binoculars, camera, and notebooks, high up into the trees, find a suitable branch, and settle in—trying not to look down—for five or more hours of waiting for "her" flowers to be visited by their pollinators. She might slide along the branch to get closer to the action or try to collect some specimens, always aware that her only supports were one slender rope, two metal clamps, and a moss-covered, slippery branch.

Because rainforest trees are so long-lived, it is impossible for a single scientist to study the dynamics of a species over its full life, which may range from 150 to 1,400 years. In fact, we don't even know how long rainforest trees do live. Because the tropics do not have distinct seasons, the trees don't form clear annual growth rings. "There are virtually no published reports of growth rates for trees in natural forests of the tropical lowlands," Gary Hartshorn, of Costa Rica's Tropical Science Center, told me. According to Paul Richards, the British botanist whom many would consider the grand old man of tropical rainforest botany, scientists have measured the average maximum lifespan of only two species, *Shorea leprosula* (250 years) and *Parashorea malaanonan* (200 years).

As the surprisingly complex life cycle of the brazilnut illustrates, each plant and animal in the rainforest depends upon many other species. Because virtually nothing is known about the vast majority of these complicated interdependencies or about the breeding and feeding requirements of most species, we cannot know whether setting aside small, isolated patches of forest will enable those species to survive. The Brazilian government and the World Wildlife Fund are cooperating on a twenty-year project to see how wildlife survive in reserves of various sizes, from 2.5 to 2,500 acres. The aim is to discover how big a rainforest reserve must be in order to preserve the characteristic diversity. It is, according to Tom Lovejoy of the World Wildlife Fund, the largest controlled experiment in the history of ecology. It will take at least twenty years to gauge the effects of being cut off from the larger forest on the bigger reserves. But already it is clear that the very small plots are worthless. "They deteriorate within just months," said Lovejoy. There is a body of feeling among tropical ecologists that to be sustainable, rainforest reserves should be at least 125,000 acres in extent—much bigger than many existing reserves; some say the minimum viable size is 500,000 acres.

The task of simply naming the plants and animals of the forest has hardly begun. "The tropical forest floras of Oceania are not very well known . . . and those of the Americas not at all well known," according to the United Nations 1978 overview of the subject. "There are many species of which only the genus is known (and sometimes only the family) and many other plants only known by their vernacular names—many of which refer to more than one species." In the Brazilian share of the Amazon Basin, for example, the largest area for which there is both reliable and published tree survey data is a mere fourteen acres. Botanists looking at one patch of rainforest near Manaus in Brazil, found 1,652 trees and plants, including 100 that were entirely new to science. Twenty of those had no common names, an indication that even local people knew little of them. In Europe or the United States an experienced botanist or zoologist walking through a forest, and equipped with one of the many species catalogues that exist for every temperate ecosystem, could expect to recognize practically all the plants or animals along the way, something no tropical scientist could do. Identifying rainforest trees, for example, calls not only for an encyclopedic memory but also a new way of looking at them. Ordinarily botanists distinguish between species by their flowers, but in rainforests, where trees flower at irregular times of the year and flowers blossom far above, out of sight, botanists have to learn to distinguish between various patterns of bark and minute differences in the texture and arrangement of leaves, a rare skill even among the most experienced tropical botanists. Few scientists ever become as expert as natives of the forest in distinguishing between the many hundreds of forest species.

The tropics have more species than temperate regions, but fewer specialists, and fewer and poorer facilities. There are practically no comprehensive listings or collections of species for a particular country or region. Brazil, which is larger than the continental United States and is one of the richest and most developed of the tropical countries, has forty-nine plant collections, but they contain only two million specimens combined, compared with forty-six million in the United States. The Royal Botanical Gardens at Kew, in England, has better collections of the plants of many rainforest countries than those countries do themselves.

Even the questions of how much rainforest is left and how fast it is being destroyed are unanswered in some countries and answered inadequately in many others. Satellite sensing, radar, and aerial photographs give accurate information about forest cover, but many countries use official figures based on out-of-date surveys that were never more than crude estimates. The Indonesian government, for instance, has for years claimed that the country has 305 million acres of rainforest remaining, despite the fact that at least 1.5 million acres are deforested—and a much larger area degraded—each year. The amount of unspoiled rainforest in Indonesia may be as low as 200 million acres. Uttarakhand, a Himalayan district in the Indian state of Uttar Pradesh, is according to official figures 67 percent covered by forests. Satellite images, however, reveal that the true figure is 37.5 percent. When Norman Myers, a British tropical ecologist, compiled figures on tropical deforestation for the U.S. National Academy of Sciences, in the late 1970s, he had no response from Zaire, a country that is believed to contain one tenth of the world's rainforests, to any of his thirty-two inquiries.

Fundamental tropical biological research (of which rainforest research is only a part) gets about $40 million a year worldwide. Half the research money is from American sources, although the United States and its territories contain less than 1 percent of the world's rainforests. Despite this apparently generous contribution, some American scientists want the United States to boost spending on tropical ecology. They point out that the National Aeronautics and Space Administration (NASA) gets more than $5.5 billion a year, even though the United States owns none of outer space.

There are at most four thousand scientists in the whole world who are primarily concerned with tropical ecosystems, according to the National Science Foundation. About half of these are taxonomists, whose concern is simply to name new species. One half of all tropical ecologists and taxonomists are North Americans and Europeans (17 percent in Europe, the rest in North America). Latin America has 22 percent, as do Asia and Australia together; Africa has 6 percent. The total number of scientific papers published each year on the environmental biology of the United States alone is greater than the total published on all aspects of tropical biology worldwide. "Only a handful of individuals in the United States is competent to supervise and undertake large-scale studies of tropical ecological systems involving experimental modification," says Peter Raven, chairman of the National Research Council's committee on research priorities in tropical biology, a part of the National Academy of Sciences. "And no more than three of these individuals are presently engaged in tropical studies. Worldwide the total of scientists competent to undertake such studies amounts to no more than two dozen."

5

Living in the Forest

Only a few people, the traditional forest dwellers, can survive in the rainforest without damaging it. They grow food on its infertile soils. They hunt and fish without driving the wild creatures of the forest to extinction. They know tens of thousands of edible, medicinal, and poisonous plants. And, more effectively than any "modern" society, they limit their own populations to what their world can support.

As roads open their ancestral lands to prospectors and settlers, these long-isolated groups suffer cultural and even physical extinction. They have no immunity to the diseases brought by outsiders. Their defenses are nothing against the guns of those who usurp their lands. The legal recognition of their natural rights is vague, inadequate, and unenforced. They come under intense pressure from government agents and missionaries to abandon their languages and their ways of life and to blend in with the dominant society, a society they can only enter at the lowest level, as itinerant laborers or beggars.

In 1542 the Dominican priests Bartolomé de las Casas and Antônio de Montesinos proposed a novel concept. They suggested that tribal people are human beings rather than subhuman or bestial. This was accepted by Pope Paul III, who pronounced sentence of excommunication against anyone who deprived Indians of their liberty or property. Philip IV supported the pope by declaring that any Spanish subjects accused of those crimes would be judged by the Inquisition. This view of matters did not catch on among the wider public, however. In 1972, for the first time in Colombian history, a group of white farmers were brought to trial for murdering Indians. Among the sixteen Cuiva Indians killed were seven children. The Supreme Court acquitted the farmers on the grounds that they did not know they were acting wrongly because they did not think their victims were human, and the laws protecting them were relatively recent.

Even today, the Ayoreo Indians, the last nomadic group in Paraguay, are being hunted down by American missionaries. The missionaries belong to the Florida-based New Tribes Mission, an evangelical organization that operates in sixteen countries and has an annual budget of more than $16 million. They fly over the forest in light airplanes, searching for isolated Ayoreo settlements. When they find a group, the missionaries send out parties of what they call "tame" Indians (that is, those who have been converted to Christianity), frequently armed, to bring the uncontacted forest Indians back to the mission. Often the hunters and the hunted, though both Ayoreo, are traditional rivals, and on occasion there has reportedly been violent conflict between the Christians and their quarry. A letter dated March 1979 from Lane McClure, one of the New Tribes missionaries, to his "prayer partners" in the United States describes one manhunt. In January of that year, twenty-seven armed Indians from the El Faro Moro mission post captured twenty-four "wild" Ayoreo, including four children.

> When the El Faro men were close they started shouting their names and that they had come in peace. The "pig people" [what the missionaries called this group of the Ayoreo] men began shouting also as their women ran to the safety of the "monte." Tension ran high as these two arch enemies stood looking at each other. . . . The turning point seemed to come when Cadui, one of the El Faro men, threw his rifle behind him and walked forward. There were 12 men and 8 women. However they had to wait three days before all the women were rounded up; they were scared to death.

Once in the camp, many Ayoreo suffer, and some die, from shock, inadequate food, exposure to diseases to which they have no immunity, and lack of medical care. Linda Keefe, a New Tribes Mission nurse, told Luke Holland, who now works with Survival International, a group that campaigns for the rights of tribal peoples, "The newcomers don't like the food we give them so they get real skinny. They want their jungle food, but we don't have a whole heap of jungle food lying around." When Holland asked her why Ojoide, the leader of the group whose capture was just described, died within a few weeks of being brought to El Faro Moro mission, Keefe said, "The newcomers have got a kind of flu—it's going round now—but these guys don't get over it so quick. . . . He was so full of antibiotics I can't see how he could have died." She added, "They say we're running a concentration camp, but one of the missionary wives said, 'I'll bet anybody living in a concentration camp would want to change places with one of our Indians.' " Although the New Tribes Mission has denied the overall charges against its activities in Paraguay, it has never responded to Survival International in detail.

Acts of high-principled cruelty such as these do not spring from personal hatred. In most cases they are ultimately bound up with economic and political forces. In 1959 an American company, Pure Oil, found gas deposits in the Ayoreo's traditional lands. In the same year, to clear the way for the company's prospectors, the government of Paraguay gave the New Tribes Mission six thousand acres of land and the right to move the Ayoreo there and "civilize" them. The Ayoreo's chances of being allowed to return to their homelands were not improved by the recent discovery of uranium deposits in the region. An American company, Teton Exploration and Mining, is now prospecting there for uranium.

. . .

Aborigines are the original inhabitants of an area. In the rainforests of western and central Africa the Pygmies and the Bushmen were the aborigines. The Bushmen have now disappeared completely from the forest; they live mainly in the Kalahari Desert. There are perhaps 200,000 Pygmies left in the African rainforest, but most of them have mixed with the dominant Bantu culture. About 25,000 Pygmies of the Mbutu group, living in the Ituri Forest in eastern Zaire, have not intermarried and have retained their traditional culture.

Pygmies are nomads who live simply, with few possessions, but other groups who came later to the rainforest established extensive and highly organized cultures and trade networks. In the fifteenth century, when Portuguese explorers reached the coast of central Africa, the kingdom of the Congo commanded a four-hundred-mile stretch of territory along the river, from the mouth of the Congo up to what is now Kinshasa. Other highly organized African rainforest cultures included the Ashante of Ghana, whose beautiful gold jewelry and symbols of office are now dispersed to museums far from the Ashante homeland. The Bantu people, who now dominate the African rainforest, are comparative newcomers, having lived there for only 1,500 to 2,000 years. They have had less time to adapt to the forest, one result being that the African rainforest has been more altered than the Amazon forest or large areas of the Asian forests.

Identifying the aboriginal inhabitants of the Pacific islands and Southeast Asia is difficult. There has been such a cross-fertilization of populations among islands and between islands and the mainland that it is impossible to say who the original inhabitants of any area were. We do know, however, that humans have been here for at least the past forty thousand years and possibly much longer. Many groups have been living in the forest for thousands of years, and farming and hunting techniques have been modified gradually as various groups mingled and intermarried. There are signs of human settlements in rainforests nine thousand years ago in India and New Guinea. The earliest such signs in Africa are three thousand years old, and in Latin America seven thousand years old.

No one knows how or when the original inhabitants of the South and Central American rainforests got there. One theory is that during the last ice age, when the sea level was lower, hunting parties tracking mastodons crossed over from Siberia via a landbridge where the Bering Strait now is. About ten thousand years ago, people for the first time reached the lowland forests of the Amazon Basin. They found a difficult and in many ways hostile environment. Originally they survived as hunters and gatherers, living in small, nomadic bands. Over the centuries, however, they adapted so well to the forest, physically and culturally, that many areas in South and Central America developed dense, highly organized populations, with sophisticated means of production and extensive trade networks. The second wave of immigrants to South America, the Europeans, remarked on the crowded margins of the Amazon River. Father Cristobal de Acuña, the Portuguese priest who wrote the first published account of Amazonia (1639), noted: "These nations are so near each other that from the last villages of one they hear the people of the other at work." All the aboriginal cultures of the floodplain were completely destroyed within 150 years of their first contact with Europeans.

Anthropologists estimate that in 1500 the Amazon Basin had a thriving population of from 6 to 9 million. In 1900 Brazil had one million Indians. Today there are fewer than 200,000. About half the 230 tribes that lived in Brazil at the turn of the century are extinct. Among the forest groups affected during that period were the Nambiquaras, who declined from 10,000 to less than 1,000; the Tembes and Timbiras, whose population dropped from 6,000 to less than 60; and the Kayapos, of whom there are fewer than 10 left of a population some twenty years ago of 2,500. Eighty-seven tribes, each with its own distinct culture, have been completely exterminated in Brazil since the turn of the century.

There are perhaps a thousand or more forest tribes around the world—many if not most on the verge of extinction—and doubtless some that are still unknown to the outside world. In the 1970s, while Brazil was building the Trans-Amazon Highway, previously unknown groups of forest dwellers were encountered at a rate of one per year. Colombia has sixty tribal groups. There are seven million tribal people comprising eighty distinct cultural-linguistic groups in the Philippines. Indonesia has 360 distinct ethnic groups, many of whom speak only their tribal language. Two hundred tribes live in the Congo Basin. Papua New Guinea has more than seven hundred tribes, mostly forest dwellers, each with its own distinct language.

The subtlety and depth of the adaptations of aboriginal people to their ecosystems is clear in the difference between Pygmies, who have lived in the forest for many millennia, and Bantu-speaking groups, who came much later. Pygmies get a good living from the forest—hunting, fishing, and collecting plants—without too much hard work. They know how to move through the forest without fear or danger and can set up their camps in a single morning. They have plenty of time to tell stories and to teach their children about the forest. Bantu farmers battle against the forest. Clearing land for their garden plots is backbreaking work; building villages takes months, and once the elaborate settlement—requiring large treeless spaces—is complete, it is hot and dusty and a mecca for bothersome and dangerous insects. Nonetheless, the Bantu are more powerful than the Pygmies, whom they try to bind into their service. They regard Pygmies as their slaves but depend upon them for honey and other forest goods because the Bantu are afraid to go deep into the forest. Pygmies encourage this fear, since it gives them the freedom of the forest, where they are independent of Bantu control.

The gap between Pygmies and Bantu is as nothing compared with the gap between forest people in general and outsiders. Forest people are physically adapted to the hot, humid environment in which they live. Their metabolic rate (the rate at which chemical processes take place within their bodies) is lower; therefore forest people generate less heat than do Europeans or North Americans. In addition most aboriginal people are relatively short and small, which gives them a large surface area in relation to their bulk. This means that heat can dissipate from their bodies more quickly. As a result, forest people sweat less and need less water than do outsiders. This is just as well, because in the humid atmosphere of a rainforest, where evaporation is low, sweating is an inefficient way of cooling down.

Forest people do not in general suffer from cancer or the chronic degenerative diseases, such as hypertension and heart disease, of industrialized societies. Stress and obesity are virtually unknown in unacculturated forest groups. The overall health of deep forest dwellers, those whose environment and way of life are relatively uninfluenced by the dominant society, is good. Adults, at least, show few signs of nutritional deficiency. Nonetheless, the life expectancy of most forest people is perhaps half that of a middle-class European or North American, who can stay alive for many years in an unhealthy state. When they cease to be healthy, forest people die.

. . .

It is difficult to generalize about how well nourished forest people are. Conditions vary from community to community, and a great deal depends on the degree of influence of the dominant society. One study, by Rebecca Holmes of the Venezuelan Institute of Scientific Studies, suggests that the difficult conditions of life in the forest take their toll among the children of a community. In two relatively isolated groups of Yanomama Indians she found that "most Yanomama children from one to twelve years old are relatively thin, many falling in the moderate to severely malnourished category." Beyond the age of puberty, however, the Yanomamas have a better-than-average nutritional status, as compared with reference populations in England and Caracas. It thus appears that the good health of the adult Yanomama Indians may be the result of a process of natural selection that weeds out the unfit before they reach adulthood.

But data on the nutrition and health of forest communities is difficult to analyze, especially since it now seems that some of our notions about health and nutrition may not be appropriate to other cultures. One such notion is that taller and bigger people are better nourished and healthier than short, thin people. By this standard, the short stature and slight build of most forest people are signs of inadequate diet. But, says a 1978 UNESCO study, "It is not certain that the increases in height and weight of children and, presumably, the increased height and weight of the adult population provides only benefits to the recipient. Metabolic and degenerative diseases seem to be much more prolific in well-nourished communities."

Lowland tropical rainforests, because of their soil and climatic conditions, tend to produce plants with a lot of bulk but low protein content. Most forest people hunt and fish to supplement their protein intake, but few reach the levels considered adequate by Europeans and North American standards. Evidently, however, rainforest people can tolerate fairly low levels of protein intake better than can their cousins in the temperate world. One adaptive mechanism is an ability to store protein in their bodies for several weeks. This suits the life-style of people whose protein consumption is intermittent, with long periods without meat punctuated by great meat-eating feasts after successful hunts. Similarly, though outsiders who adopt a forest diet often suffer from deficiencies of essential trace nutrients—such as calcium, zinc, sodium chloride, and certain vitamins—such deficiencies are rare among forest people eating the same foods.

Detailed studies by M. W. Corden and H. A. P. C. Oomen of more than two hundred adults living in the rainforests of New Guinea found no evidence of any nutritional deficiency. Yet their diet was so low in "essential" nutrients that scientists hypothesize that they have developed special nitrogen-fixing intestinal flora to supplement food nitrogen. Another study, by S. A. Schlegel and H. A. Gutherie, in the Philippines, compared the diets and health of members of two communities of the Tiruray tribe. The "typical" male of the forest-dwelling group was a thirty-five-year-old man named Silu. Bekey, a fifty-four-year-old peasant farmer, represented the group of settled farmers. Silu's daily energy intake was 1,176 kilocalories. Bekey's was almost twice as great, 2,236 kilocalories. By accepted standards, Silu's intake is "almost impossibly low," yet he showed none of the expected consequences of his low energy diet—poor nutritional status, physical apathy, and other ill effects. Alfred Wallace was similarly struck by the eating habits of people he met while collecting new species along the Rio Negro.

> At first I was quite puzzled to find out when they had their meals. . . . I could not imagine that they really had nothing else to eat, but at last was obliged to come to the conclusion that various preparations of mandiocca and water formed their only food. About once a week they would get a few small fish or a bird, but then it would be divided among so many as only to serve as a relish to the cassava bread. My hunter never took anything out with him but a bag of dry farinha, and after being away fourteen hours in his canoe would come home and sit down in his hammock and converse as if his thoughts were far from eating, and then, when a cuya of mingau was offered him, would quite contentedly drink it, and be ready to start off before daybreak the next morning. Yet he was as stout and jolly-looking as John Bull himself, fed daily on fat beef and mutton.

In many forest groups adults show a high tolerance of infectious diseases, such as chicken pox, mononucleosis, and hepatitis B. In some groups practically the whole population has one or more of these diseases without appearing to be severely affected. Children are not so resistant, however, and infectious diseases are an important factor in the high incidence of infant and child mortality in many forest communities. Outside the forest, where such diseases are much less common, they are often fatal unless the victim receives medical treatment. Many groups of forest dwellers also appear to be immune to certain other serious diseases, such as leishmaniasis and yellow fever, that commonly strike newcomers to the forest.

The diseases that most seriously affect aboriginal populations—malaria, dysentery, typhoid, measles, tuberculosis, influenza, and others—are associated with change in their environment. Many disease-carrying insects, including those responsible for malaria and yellow fever, live in the forest canopy; only when their habitat is destroyed do they come nearer the ground and prey on humans. Dysentery and typhoid are associated with poor water quality, resulting from overpopulation or pollution. Measles, tuberculosis, influenza, and other such diseases do not exist in still-isolated groups of forest dwellers.

. . .

Cultural changes in diet and living patterns as a result of contact make the situation worse. The Pacific islands suffered a severe epidemic of influenza when the islanders were forced to wear clothes. Shortly after the first government patrols appeared in the Jimi Valley of Papua New Guinea in the late 1950s, an alarming number of babies in the area were born with endemic cretinism. It turned out that the patrols used salt to pay locals for goods and services. This "patrol salt" supplanted the salt the people of the Jimi used to obtain in trading with their neighbors. But unlike the traditional salt, the patrol salt lacked iodine. The iodine deficiency it created in the Jimi people was the cause of their cretinism.

The good health of forest people is bound up with their cultural practices and their land. When a group is forced to leave its homeland and begins to lose its culture, many protective mechanisms are lost. Many hunter-gatherer societies show, as a United Nations survey puts it, "obvious signs of freedom and absence of constraint which appear to many observers as an expression of environmental fitness and happiness." One of the most important adaptations forest people make to their environment is limiting the size and density of their populations to the level the ecosystem can sustain. If a group is below a certain size, for example, various infectious agents, including those that cause mumps, measles, and influenza, will not be able to survive. There are many ways of controlling population size, including regulating the age of marriage, sexual taboos, contraception, abortion, infanticide, death penalties, warfare, and abandoning the ill. Population density is regulated by warfare, limits on the size of extended families, and other, more abstract cultural values, such as a desire for freedom and a fear of outsiders.

Betty Meggers, in a classic study of Amazonian Indian groups, pointed out the difference in the traditional population control strategies of groups who live on the fertile floodplains and those who live deeper in the forest. Forest groups, such as the Jivaro, exist in a stable environment, but one that supports relatively few people. Villages must be small, able to move frequently, and widely spaced. In order to keep their numbers low, the Jivaro use to some degree all the measures just mentioned except abandoning the ill.

The floodplains, where each year the river deposits fresh loads of fertile silt, are good for agriculture and can support relatively large numbers of people. But during the six- to nine-month-long rainy season, while the river is high, villages have to decamp to higher ground and must live off the food they have grown and stored earlier in the year. This pattern of high productivity followed by months of shortages means that the ideal size of floodplain groups varies throughout the year. A large labor force is necessary and feasible in the planting and harvesting season but may be a burden during the rainy season. Floodplain groups such as the Omagua solved this problem by treating the captives they took in warfare with other groups as agricultural slaves that could be disposed of when food was scarce. Other population control methods they used were infanticide and warfare itself, but the slave system acted as a buffer that saved them from having periodically to kill large numbers of their own people.

Conservation measures, like population controls, are embedded in the social and religious structures of forest cultures. Religious taboos or the authority of elders often serve to conserve natural resources. Some South American Indians release captured birds after taking a few feathers for their headdresses. Headmen in the Andaman Islands protect the islands' most important wild foods from overexploitation by setting limits on the harvest from time to time. Anyone who disobeys is severely punished.

A romantic view of "primitive people" emphasizes the fragility of tribal societies and overlooks the robustness and adaptability that enabled them to survive in the difficult forest environment. That image is not only wrong but dangerous in that it suggests that no tribal culture can survive in the modern world.

Healthy tribal societies are dynamic, not immutable. Like our own, they evolve and change in response to changing conditions. The question concerning the survival of forest groups is not whether they can adapt to the twentieth and twenty-first centuries without losing their cultural identities but whether they will be allowed to do so. Cultural extinction is not inevitable. To avoid it a group must have land, protection from introduced diseases, time to adapt, and the right to determine its own future.

Some groups are fiercely independent and suspicious of outside influences; others are receptive to new ideas and eager to learn more about new things and new techniques. Each deserves the time and opportunity to make its own choices about whether to adopt elements of other cultures and at what rate. All need the twin buffers of medicine and education, if they are not to be overwhelmed both physically and culturally by their first encounters with the dominant society. They should be acquainted with the problems faced by other forest tribes who have allowed their own cultures to lapse. They need to know some of the language, the history, and the mechanics of the dominant society. They must also be aware of the changes and increased responsibilities that come with the adoption of certain innovations. The use of firearms, for example, will require new hunting patterns if the wildlife pool is not to be exhausted. Medicines and food supplies will increase population pressure and may require the construction of roads through tribal lands.

Members of the dominant society, especially those who enter tribal lands, should also know something about tribal cultures and about the importance of traditional skills and knowledge. On a purely material level, modern agronomists who have found that their own methods of farming do not work in tropical rainforests have learned a great deal from the agricultural techniques of forest dwellers. Forest farmers have developed efficient systems of erosion control, nitrogen enrichment, groundwater exploitation, slope management, fertilization, drainage, intercropping, forest regeneration, weeding, irrigation, and wildlife management—systems that are still little understood by outsiders.

Forest tribes have a holistic view of the world, the wisdom and practical value of which is just beginning to be appreciated by ecologists, economists, and a few development specialists. The now fashionable notion of integrated rural development is a step closer to the "primitive" notion of life, in which not only its physical and economic features but also its cultural, psychological, and spiritual aspects are recognized as influencing one another. Aboriginal societies are the world's oldest cultures. They are the last people on earth to live close to nature. Not only the skills, but also the wisdom, the social patterns, and the outlook of aboriginal peoples constitute a fund of knowledge that, like a genetic bank of wild plants, technological man may need to call upon in the future.

. . .

The legal rights of forest people vary from country to country. They are affected by regulations as diverse as those concerning forestry, religious freedom, mineral rights, and land ownership—and even more so by the attitudes and abilities of those responsible for tribal affairs. In Indonesia, for example, the constitution guarantees freedom of religion only for believers in "One Supreme God," a phrase that does not apply to the animist beliefs of many aboriginal groups. A century ago Colombia passed a law saying that "the general legislation of the Republic shall not apply to savages who are being brought to civilized life by the Missions." Instead the state delegated its civil, penal, and judicial authority over forest-dwelling groups to the Catholic Church. This absolute control was somewhat modified in 1974, but the Church is still the main day-to-day authority in many areas. Bolivia, a country in which 75 percent of the population is Indian, has assigned responsibility for the education, medical care, and general welfare of its forest-dwelling Indians to the Summer Institute of Linguistics, an evangelical Protestant sect.

Insofar as they have an official policy toward forest people and other cultural minorities, most governments aim to "integrate" them into the dominant culture. Mauricio Rangel Reis put this view succinctly in 1974, when as Brazil's minister of the interior he said, "We are going to create a policy of integrating the Indian population into Brazilian society as quickly as possible. . . . We think that the ideals of preserving the Indian population within its own 'habitat' are very beautiful ideals, but unrealistic." A few years ago Survival International obtained the exercise book of a thirteen-year-old Witoto Indian boy attending a Catholic mission school in a remote part of the Colombian Amazon. It included a self-portrait of the boy, who in real life eats with his hand from a communal dish, sitting at a table set with knives, forks, spoons, plates, glasses, and a vase with flowers. Laboriously copied-out essays on such subjects as How to Behave in the Street and How to Use a Handkerchief included one on theatres: "When a play is presented we should applaud, but not in a vulgar manner, nor laughing, because that shows bad education."

Conformity is convenient. Administrators find their jobs easier in a uniform society. Nomads are difficult to keep track of and tax; modern legal systems are not geared to collective land ownership; religious practices that may require the use of hallucinogens are a threat to a society in which drugs can be abused but not used sacramentally. But something more powerful than a straightforward desire for administrative convenience motivates many who work to "integrate" minorities. They want indigenous people to undergo a spiritual as well as a behavioral conversion, to change their minds as well as their clothes and their language.

The native peoples of the Mentawai Islands of Indonesia (Siberut, Sipora, and North and South Pagai) have until recently followed a complex and stable way of life, based on hunting, fishing, and shifting cultivation. Mentawai Islanders live in extended families in longhouses, or *uma*s, which are generally built near rivers. The main crops are sago palms, bananas, and taro, intercropped with trees and spice bushes. Their system of agriculture does not involve burning the forest. After clearing a small plot, they leave the debris to decay, slowly releasing nutrients and providing shade for the early crops. The forest that in time recolonizes the site is more like the natural forest than would be the case if fire had destroyed the tree stumps and the soil nutrients. These "domestic forests" are good homes for wildlife and sources of medicines, fruits, and building materials.

Sago palm, from the pith of which the Mentawaians obtain their staple food, sago flour, deserves more attention as a subsistence crop. It gives a better yield in relation to labor expended than virtually any other crop. Sago palms are self-propagating. They grow in swamps, where there is little competition from weeds. They have no natural animal enemies. Edible grubs can be encouraged to breed in the bark. Sago is a good feed for pigs and chickens, and though it is low in protein, it requires so little work to grow that cultivators have time to supplement their diet by hunting and fishing.

The Mentawai people collect forest products, but they cut only one kind of tree, the sal tree (several species of *Shorea*), from which they make canoes, their main form of transport. Hunting with bows and arrows is a social activity, surrounded by religious taboos that protect wildlife stocks. The four islands, which together are only half the size of Wales, include among their fantastic wildlife four species of primates found nowhere else on earth.

In the past sixty years, especially since 1960, the Indonesian government and missionaries have established fifty villages in the islands. The authorities discourage the traditional housing pattern, based on *uma*s. They prefer large compounds of nuclear families centered on the church. (In 1954 the Dutch colonial administration gave Mentawaians three months to choose Islam or Christianity as their new religion.) The result, as cultural population controls have been abandoned, has been a dramatic increase in numbers over the past twenty years, averaging 3 percent a year. Rice, a more "civilized" crop, is replacing sago palm, though rice requires more labor for a lower yield. There is less time for hunting and, because the villages are concentrated, the hunting area is smaller and overexploited. The traditional taboos that prevented overhunting are being forgotten. Due to the poor rice harvests, the villagers depend more and more on store-bought goods.

There is also a new system of land tenure. The people own the land, but the Forestry Department has rights over the trees and has leased more than 90 percent of the land area of the islands to big logging companies. One percent has been set aside as a nature reserve. Under the new law, the government owes no compensation to the villagers for the loss of any forest product, including the sal, which is their only good canoe tree. The logging is not only depleting an important capital resource, it is also silting up the rivers, making navigation difficult and reducing the fish catch. Few Mentawaians have jobs with the logging companies. They are not considered good employees because they have no experience working to other people's timetables and are more concerned with tending their crops.

At the request of the Indonesian government, the World Wildlife Fund and Survival International (WWF/SI) have developed a plan to protect Siberut and help the islanders there adjust to the changes in their environment and their lives. The island is to be divided into three zones: a two-hundred-square-mile nature reserve, for which the government has already canceled all logging concessions; a four-hundred-square-mile traditional-use zone; and a one-thousand-square-mile development zone. The nature reserve is to be completely protected from disruption. In the traditional-use zone the islanders will be able to pursue their traditional forest activities. Survival International hopes that new fishing techniques and the introduction of more protein-rich varieties of sago palm will help offset the reduction in animal resources that is driving them to buy food or rely on handouts.

The national government has made the WWF/SI project official policy, but it has not yet canceled logging concessions in the traditional-use zone. In fact, according to Anton Fernhout, the project manager, "the minute our full-time person left, the loggers came in again. They are in a hurry to destroy as much as possible before we get our act together." Witnesses also say that tribal people are once again being forcibly resettled away from their traditional lands in the interior, where the loggers are operating. "The military are driving them from their longhouses to the coast. They are forced to live in wooden huts and cultivate rice and give up their traditional way of living."

Cultural and physical health are so intimately connected that a change in either one ricochets back and forth between the two, gaining instead of losing force on each bounce. When Western medicine is available and succeeds where their own remedies have failed, people begin to lose faith in their institutions and feel impotent. Loss of faith in traditional medicine often leads to an increase in incurable illness, since every culture has a number of ailments, particularly psychosomatic ones, that can be cured only by the magicomedical practices of that culture.

As a group learns to depend on goods it cannot produce, and as it sees even its leaders treated with contempt by the dominant society, it becomes demoralized. With the weakening of their secular and religious authority, the elders can no longer impose the traditional controls on population and use of natural resources. As the group abandons its traditional skills, it finds making a living from the forest more difficult. The people feel helpless and dependent. They have lost control of their lives.

The cultural disintegration of the Panare, who live in Venezuela near the Orinoco River, was deliberately planned by the evangelical New Tribes Mission. An American couple named Price settled near a Panare village in 1972. The Panare were friendly to the Prices and even helped them build a house. The Indians traded with the missionaries, giving them food in return for iron tools. One New Tribes missionary described to the writer Norman Lewis the technique for making contact with Indians. "We leave gifts . . . knives, axes, mirrors, the kind of things Indians can't resist. . . . Naturally they want to go on receiving all these desirable things we've been giving them and sometimes it comes as a surprise when we explain that from now on if they want to possess them they must work for money. . . . We can usually fix them up with something on the local farms. They settle down to it when they realize that there's no going back."

All followed according to plan, but the Panare showed no interest in the Prices' biblical message. Then suddenly in the late 1970s the Panare began giving up their own culture. Maria Eugenia Villaon, an anthropologist who had lived among the Panare and recorded some of their music, returned the following year to complete a census. When she played back their tape-recorded songs, the Panare jumped back in terror and clasped their hands to their ears. In those songs, they said, when she had switched off the music, the devil was speaking through them. They sang instead for her a Panare version of "Weary of earth and laden with my sin." The Panare spoke of guilt and sin and punishment, though these concepts are foreign to their culture and their language had no words for them.

Another anthropologist, Henry Corradini, who had lived with the Panare, decided to find out how the New Tribes missionaries had succeeded where the proselytizing of Jesuits and Franciscans had failed. Reading the two books of Bible stories that the missionaries had translated into Panare, he found that the stories had been distorted to give the Panare the idea that *they* had killed the son of God.

> The Panare killed Jesus Christ because they were wicked.
> Let's kill Jesus Christ
> said the Panare.
> The Panare seized Jesus Christ.
> The Panare killed in this way.
> They laid a cross on the ground.
> They fastened his hands and his feet
> against the wooden beams, with nails.
> They raised him straight up, nailed.
> The man died like that, nailed.
> Thus the Panare killed Jesus Christ.

The missionaries also told the Panare that God would take revenge for their wickedness, unless they repented.

> God will exterminate the Panare
> by throwing them onto the fire.
> It is a huge fire.
> "I'm going to hurl the Panare into the fire," said God.
> God is good.
> "Do you want to be roasted in the fire?" asks God.
> "Do you have something to pay me with
> so that I won't roast you in the fire?
> What is it you're going to pay me with?"

. . .

Dirty camps full of listless, emaciated people wearing cast-off T-shirts and torn dresses, people whose eyes tell of outrage and defeat, do not create the kind of image governments like to project to foreign visitors. For these it is sometimes convenient to keep forest groups in a pristine condition, by force, if necessary. *Flying Colors,* the

Braniff Airlines in-flight magazine, in one of the last editions before the airline folded in 1982, featured an article about Ecuador. "Excursions to an Auca Indian tribe can be made on Wednesday or Thursday from Quito. The limit of the jungle aircraft restricts the size of the group to 15. Observe Auca daily activities. Watch them cook meals over an open fire, weave, and make darts with which to hunt birds and monkeys."

PANAMIN, the agency charged with enforcing government decisions affecting tribal minorities in the Philippines, has a number of reservations where "enforced primitivism" is the rule. There indigenous people are forbidden simple technologies that could make them self-sufficient, so that tourists can see them in a "pure" state. The members of the board of PANAMIN have extensive financial interests in the land of the people whose interests they are pledged to protect. Its chairman, Manuel Elizalde, Junior, scion of the Philippines' fifth richest family, has mines, sugar plantations, distilleries, and other industries based in tribal lands. A PANAMIN employee said, "What made me really doubt Elizalde's interests in the minorities was that, on our advance visits, I was always accompanied by two prospector-geologists, a Russian and a Filipino. . . . I found out later that they are hired by Elizalde on a commission basis." A Mansaka tribal leader addressed the agency on behalf of his people: "We do not understand PANAMIN's aims. It gives us money to attend meetings; to perform in their cultural shows for tourists. It promises us many things but does not fulfill them. Is PANAMIN for the natives?" His question was answered indirectly when Elizalde told a tribal meeting in 1975, "I am doing this for your own welfare so that you will be able to send your children to school, have them graduate, get a good job and enjoy good whiskies on the mountains."

. . . .

Land is the most important element of any forest tribe's existence. Land—in fact, the entire physical environment, the ecosystem—has an important spiritual meaning for virtually all forest groups and a sacred connection with the culture itself. Once deprived of their land, forest dwellers can only look forward to being absorbed into the lowest level of the dominant culture, as landless peasants, low-paid laborers, or beggars. The Conibo Indians of Peru are a case in point. They now work in lowly positions for the logging company that has taken control of their ancestral lands and is cutting down the forest that for centuries provided the Conibo with a home and a way of life.

Land laws vary enormously from country to country. There are a number of international agreements concerning the rights of tribal peoples to their traditional lands. Twenty-seven states have ratified the International Labor Organization's convention on tribal and indigenous peoples. Under it governments are bound to adopt "special measures for protection of the institutions, persons, property, and labour of indigenous populations." Article 11 states that "the right of ownership, collective or individual, of the members of the populations concerned over the land which they occupy shall be recognized." The American Declaration of the Rights and Duties of Man and the Universal Declaration of Human Rights also provide for the protection of the rights of all sectors of a national population.

Papua New Guinea recognizes, in law and in practice, the right of indigenous people to their traditional lands more than any other country, saving, perhaps, New Zealand. Ninety-seven percent of the land in Papua New Guinea is in customary ownership, not owned by individuals but by tribes and clans. Peru guarantees the integrity of the land of forest-dwelling native communities, but it also allows free access for oil and gas pipelines and mineral exploitation with no compensation to the occupiers of the land. Neither Ecuador nor Bolivia, despite having signed the UN Convention on Tribal Peoples, recognizes the right of Indians to their traditional lands, nor has either country established any reserves where Indians can live as a community. Individual Indians may own land, but there is no mechanism for assigning collective ownership, a notion crucial to most Indian cultures. In the Philippines conflicting laws guarantee tribal groups rights over ancestral lands and at the same time declare that the government has a right to convert all such lands to other uses. The main relevant law is presidential decree 410. Since its promulgation in 1974, not one tribal group has had its rights over its homelands officially recognized. The general pattern is for governments to give weak lip service to the notion that tribal people should have some protected land of their own but to ignore this principle in practice and even pass laws that contradict it.

Land is at the center of almost all disputes between tribal people and industrial societies. When President Marcos issued the country's main tribal land law, presidential decree (PD) 410, the list of provinces in which cultural minorities' land might be recognized omitted Abra, the home of the Tinggian tribe. Over the centuries, the Tinggians have created some of the finest terraced and irrigated rice fields in the world. They also practice a rotation agriculture in the mountains and hunt and fish in the forest. Eleven days after PD 410 was announced, a logging and pulp concern called Cellophil Resources Corporation was awarded 500,000 acres of forestland, much of it belonging by tradition to the Tinggians. The concession also sprawls across the mountains of four other provinces and covers the land of four minority groups besides the Tinggians. It extends over four supposedly protected watersheds on which the National Power Company plans to build at least nine major dams. One of the affected watersheds is the Agno, where siltation is already choking the Ambuklao Dam. One of the areas in the concession is Balbalasang, a national nature reserve.

Cellophil was at that time a subsidiary of the Herdis group of companies, headed by Herminio Disini. Disini is related by marriage to Imelda Marcos and is a golfing partner of the president. Marcos is widely reputed to have a large personal stake in Cellophil. Since the company's entry into Abra, its officials have been guarded by members of Marcos's special Presidential Guard Battalion. Disini gained an international reputation as the middleman for Westinghouse in the negotiations to decide a builder for the Philippines' first nuclear reactor. *Time* magazine estimated that Westinghouse paid Disini $35 million for securing Marcos's approval of the Westinghouse contract. But the publicity—and Marcos's desire to distance himself from Disini—led to the government's taking over Herdis's share of Cellophil. Subsequently, the government hired Herdis to manage the company, which is now a joint venture between the Philippine government, three Japanese companies, including Mitsubishi, and the Swiss multinational Bauminter Corporation. A consortium of European banks is financing the scheme. Japanese, American, and European companies are planning, building, and installing the equipment. Spie Batignolles, the French firm that equipped Cellophil's pulp mill, supplied the wrong machinery. As a result the river, formerly the main source of water, became unfit to drink or even to bathe in. Eventually the pulp mill broke down completely, so the company can no longer process the wood it harvests. Instead it has stepped up its logging and received a presidential exemption from the nationwide ban on exporting raw logs.

Despite its policy that it "will not assist development projects, forestry or otherwise that knowingly encroach on traditional territories of tribal people without their full and voluntary consent," the World Bank is financing, on Tinggian land, a forestry plantation that will help satisfy the future needs of Cellophil's pulp mill. The plan stipulates that the local people (whom the Philippine government and the Bank say are squatters) will be moved. Some will be employed in the plantation program. They will be given a quarter-acre plot to farm for themselves, but only for as long as they continue in the program.

The Tinggians have tried to fight the takeover of their lands by pointing out certain illegalities. Cellophil's 500,000-acre concession is twice as large as government regulations allow, and a regulation that predates martial law excludes tribal lands from all concession areas. Leaders of all the tribes affected by Cellophil met in 1978 and signed a resolution asking Cellophil to stop its operations until certain changes were agreed upon. The tribal leaders wanted the company not to float logs in the river, an act that damages fish traps and irrigation canals; not to cut trees along certain creeks and watersheds where erosion would result; to set aside a number of small communal forests and pasturelands; to seek the approval of local people before undertaking any new activities. Instead, the government withdrew the people's rights to the wood in communal forests and informed them that they were "squatters" and had no legal rights.

In 1979, five hundred tribal leaders wrote to Marcos asking that Cellophil's license be withdrawn. One elder said, "We are poor. All we have now are the mountains, trees, rivers, and especially our freedom. All these the Cellophil is threatening to take away from us." In answer, the military stepped up its presence in the area. Army officers escorted Cellophil officials to meetings with local people to demonstrate the government's support for Cellophil's position. Roads throughout the province have military checkpoints to prevent mass meetings from taking place. The national government has dismissed five local mayors and put in its own men.

Many people have lost their livelihood completely, since they can no longer farm land claimed by Cellophil, and fishing, their main alternative livelihood, has been severely damaged by siltation from the clear-cutting of the forest and pollution from the pulp and paper mill. The river water, formerly the main source of drinking water, is no longer potable. Massive deforestation to feed the mills has led to erosion, landslides, and floods. Many people have been forcibly resettled far from their homes in government-run camps, modeled on the "strategic hamlets" set up by the Americans during the Vietnamese War.

Many Tinggians have been imprisoned without trial for their refusal to accept the company's presence on their land. In May 1983 Philippine Constabulary Troops of 623 Battalion entered five Tinggian villages and interrogated villagers about their alleged support for the New People's Army, the communist guerillas who oppose Marcos's regime and who are organizing much of the opposition to Cellophil. According to the Catholic Church and human rights organizations in the Philippines, the soldiers killed at least three people, including a four-year-old girl and her pregnant mother, and tortured many others. After the attack hundreds of villagers fled into the forest, abandoning their rice harvest—their only reliable supply of food for the year to come.

. . .

The will to resist extinction varies from group to group. Some, like the Txukarramaes, are fighters. The Txukarramaes live on the Xingu River in the Amazon, an area that the Brazilian government set aside in 1961 as a reservation to safeguard fourteen Indian tribes. They cooperated with the authorities in staying within the boundaries of the park, even though that meant abandoning some of their traditional territory.

In 1971 officials announced that a highway was to be built through the northern part of the park, splitting in two what was left of the Txukarramaes' land. The tribe divided into two groups. One, led by a chief named Krumari, settled near the road. The other, led by chief Raoni, were persuaded by the authorities to move south, deeper into the park. In November 1973 Krumari's band was struck by an epidemic of measles brought by the highway workers. The following month workers clearing land for cattle ranches along the highway introduced another epidemic. Many of the group died. More than seventy were taken to a government hospital, but there were no medicines to treat them. "We are out of antibiotics," one doctor told newspaper reporters, "and we urgently need vitamins, analgesics, and anti-fever drugs. The Indians are arriving with measles and broncho-pneumonia, and are in a critical state of malnutrition."

Raoni's band, and others in the park, have had a constant struggle to keep highway workers, farmworkers, and squatters off their land. They have repeatedly asked the government and their "caretakers," the National Indian Foundation (FUNAI), to demarcate their land and protect it from settlers, but without result. In a number of incidents in 1980 thirty young Indian leaders were killed while walking their own lands by settlers who were frightened or who were guarding their newly claimed property. On August 8, 1980, Raoni's warriors struck back. They killed eleven laborers who were clearing the forest for Luis Carlos Silva, a wealthy businessman who had decided to start a cattle ranch on Txukarramae land. In June Raoni had asked the men "to stop pulling down the forest and leave the area." Only days before the killings Raoni had given a public warning: "From now on anyone who trespasses on Indian territory will die . . . we will fight and maybe we will die, but we are dying slowly anyway."

Raoni claims that the attack was not meant to end in bloodshed. His nephew said later, "They should have killed the farmer not the workers, but he does not live on the farm." The massacre seems to have discouraged the Txukarramaes more than it did their foes. Afterward, emerging from yet another fruitless meeting with the head of FUNAI, a depressed Raoni said, "I think the whites should kill us all right away and take our lands and be done with it." Silva, whose hired hands were killed, commented: "The United States solved this problem with its army. They killed a lot of Indians. Today everything's quiet there and the country is respected around the world."

. . .

The most successful indigenous groups are those who master the weapons of communication, with other tribal groups and with society at large. In 1980 indigenous people in Brazil formed a new national organization, the Union of Indian Nations, with the objective of promoting cultural autonomy and self-determination, fighting for the recovery of land and resource rights, helping communities in designing and carrying out development projects, and informing the general public about indigenous peoples. Around the world many tribal groups are defending their cultural integrity more effectively, having profited from the experiences of others. A group of Ayoreo living in Bolivia, cousins of the Ayoreo being hunted down in Colombia by the New Tribes Mission, have made a film documenting their own situation and encouraging others to keep their culture and resist the expropriation of their land. The film ends with these words: "The Ayoreo have entered a new phase. Now we are no longer frightened of television, radio, and the papers. We use them and know that we can protest. We are Bolivians too and we are seeking allies."

The Kuna of Panama are among those very few tribal groups who have adapted energetically to modern ways while maintaining their cultural identity. In 1930, following a Kuna uprising, the Panamanian government formally recognized their traditional territory, the *comarca* (region) of San Blas, and signed a treaty forbidding non-Kunas from owning land there. Since then the Kuna have become the most politically powerful indigenous group in the country, perhaps in all of Central America, with their own governor, Kuna representatives in the National Assembly, and an internal congress that meets to rule on developments affecting the Kuna population. The Kuna tradition of cooperative labor continues strong. Kunas who work outside the *comarca* have their own trade unions. In the *comarca* they have wrested control of tourism from outside hoteliers.

The Kuna are using their power to turn a controversial United States–financed project to build a road through one of Panama's last remaining areas of virgin rainforest into an imaginative development that they will control. The road, largely paid for by the U.S. Agency for International Development (AID), will be an extension of the Pan-American Highway. It will run from El Llano, a small settlement on the highway, through the *comarca* to the Atlantic coastal town of Carti. Although the Kuna opposed the road in 1950, when it was first mooted, their feelings are more ambivalent today. Right now the only way to get from the town of San Blas to Panama City is by plane or by an eighteen-hour boat and train journey. The road will make trade with the capital much easier, but it will also open up the *comarca* to wealthy speculators and land-hungry peasants. Already settlers along the section of the road between El Llano and Nuevo Santeno on the border of the *comarca* have cleared virtually all the forest for several miles on either side. According to an internal AID report, many of the people to whom the government has given title to land along the road are far from being the poor farmers whom the project is supposed to help. Among the beneficiaries are a doctor who lives in Panama City, a captain of the National Guard, and a rich horse breeder.

Fearing that they would lose control of their *comarca* to settlers if the road went ahead as planned, with no controls, the Kuna took matters into their own hands. The Kuna congress has already sunk $36,000 of its own money into an agricultural settlement at the point where the road enters the *comarca,* so that the land will not be open to outside settlement. Problems stem from the fact that the ecological conditions are quite different from those to which they are accustomed on the coast. Scientists from Costa Rica's Center for Tropical Agriculture are carrying out biological inventories and land-use studies to help the Kuna find the best ways of using the newly accessible forest.

The Kuna also intend to create a botanical park in the newly opened region. The trees will be labeled with their Kuna, Spanish, and scientific names, making a sort of sylvan classroom where Kuna elders can pass on their knowledge about the forest plants and animals to the younger generation. This area, in which both hunting and clearing would be prohibited, would also play a traditional role as a place for spirits to rest undisturbed. The Kuna also want to expand the tourism business they now run on the coast to embrace scientific tourism in the heart of the rainforest. Not only would this be a source of income, Kuna leaders reckon, but travelers familiar with the beauties and fragility of the *comarca* environment could be an important source of support in any battles over sovereignty and uncontrolled development the Kuna may one day have to fight.

6

Cattle in the Clouds

One reason that the Central American rainforests seem doomed to disappear is that their destruction takes five cents off the price of an American hamburger. The United States buys up three quarters of all Central American beef exports. In 1979 this amounted to 113,000 tons, which is only about 14.5 percent of all U.S. beef imports and less than 2 percent of the beef consumed in the country. Yet producing this relatively minor contribution to the American larder has had devastating consequences for Central American forests and for Central American diets.

Costa Rica is a case in point. In 1950 cattle ranches covered 1,625,000 acres, one eighth of the country. Today over one third of Costa Rica has been converted to pastureland, though most of it is now worthless and has been abandoned. The country produces three and a half times as much beef today as it did in 1961, and a larger proportion of its production is exported. In 1979 about two of every three cattle raised in Costa Rica were exported, two thirds of them to the United States. But, since 1960, per-capita beef consumption in Costa Rica has fallen by more than 40 percent. By 1978 the average Costa Rican was eating only twenty-eight pounds of beef a year; the average American pet cat ate more.

The situation in Costa Rica is echoed throughout Central America. Everywhere pasture is spreading at the expense of rainforests and of land that could support arable crops. As a whole, about two thirds of Central America's farmland is now devoted to cattle. Peter Raven testified in 1980 before a U.S. congressional subcommittee that since 1960 more than one quarter of all Central American forests has been destroyed to produce beef. As in Costa Rica, the citizens of Nicaragua, El Salvador, and Guatemala eat on average less beef now than they did in 1961, even though beef production in those countries has tripled.

The reason for this trend is the average American's growing appetite for beef and his ability to outbid the citizens of other cattle-producing countries in order to get it. In 1960, when the annual per-capita beef consumption in the United States was 85 pounds, the country imported almost no beef. In 1978 the average American chewed his way through 122 pounds of beef. Ten percent of that, 970,000 tons, had to be imported.

A key attraction of Latin American beef for Americans is its price. It is less than half as expensive as domestically produced grass-fed beef. Indeed, in recent years the Central American beef import quotas have been an important tool of U.S. economic policymaking. The government has encouraged more imports as an antidote to the rapidly rising price of domestically produced beef. Cheap Central American imports are thus essential in the battle against inflation. Beef import quotas were increased by 13 percent in 1978 and 1979; the U.S. government estimated that this would lower the cost of a hamburger by a nickel.

The grass-fed beef of the tropics is too lean for Americans, who have become accustomed to the marbling of grain-fed cattle. So imports are usually mixed with U.S. cattle trimmings and sold for hamburgers as "American grade" meat. Much of this goes to fast-food restaurants. The rest is made into hot dogs, processed meat, and pet food. The United States government bans imports of chilled or frozen beef from Brazil because some of its cattle have hoof-and-mouth disease, but even so, because cooked meat is acceptable, over 23,000 tons of Brazilian sausages and corned beef do enter the country annually.

For some years conservationists have tried to focus attention on this phenomenon and the harm it is doing. A boycott of McDonald's, the giant of the fast-food chains, has often been mooted but never got off the ground. Although other fast-food chains, including Burger King, Jack-in-the-Box, and Roy Rogers, have admitted buying foreign beef, McDonald's management vehemently denies doing so. The company insists that the three billion hamburgers it sells each year—equivalent to 300,000 head of cattle—are made entirely of domestic beef. It is virtually impossible to disprove this claim because once Central American beef imports have been inspected by the U.S. Department of Agriculture, they are allowed to enter the domestic market without any indication of their national origin.

Before the revolution in 1979 Nicaragua was the main supplier of Central American beef to the U.S. market. President Anastasio Somoza himself was one of the biggest middlemen. Every year he sent $30 million worth of beef into the U.S. through his six Miami import companies, much of it from his own ranches. Before their overthrow the Somoza family owned one quarter of all Nicaragua's arable land. This is a pattern repeated throughout the continent. Figures from the Food and Agriculture Organization of the United Nations show that in Latin America as a whole, 93 percent of the arable land is owned by 7 percent of the landowners. That statistic does not include the millions of peasants who own no land at all.

"The majority of Costa Rican farmers, maybe 80 percent, are agriculturalists, but the cattle ranchers control much more of the land than the farmers," according to Gerardo Budowski, the director of the Tropical Agriculture Research Center (CATIE) in Turrialba, Costa Rica. The cattle industry in Costa Rica is dominated by about two thousand ranchers, who control half the agricultural land in use. In Guatemala 2.2 percent of the population owns 70 percent of the agricultural land, and devotes most of it to growing beef, coffee, and bananas for sale abroad. This inequality is what makes it possible for Americans to buy cheap beef from Central America.

"The cattle people are very powerful and they'd rather have their cattle on good land," says Budowski, whose institute tries to make subsistence farming more productive. "They'll take the poor land too—but if they have the choice they'll take the good land. Very often you'll find—and perhaps Colombia is the best example—that the best alluvial lands down in the valleys are employed for cattle raising, with very few people living on that land, and a tremendous number of poor people living on the slopes above with very poor soil.

So on the good land which could support a large population you have the rich cattle owners and on the steep slopes which should be left in forest you have the poor farmers."

Ranching as practiced throughout Central America is wasteful of land. A great deal of it is undertaken as a sort of prestige hobby by absentee proprietors, uninterested in putting time or money into the operation apart from the bare minimum needed to turn a profit. Unlike their North American or European counterparts, who have to ranch intensively because of the high cost of good land, few if any Central American ranchers use modern techniques of soil conservation, disease control, pasture management, or livestock breeding. The fact that productivity falls off sharply within a few years as toxic weeds invade the pastures, or that stocking rates average one animal per twelve acres, or that the land is simply not suitable for long-term ranching, is offset by the ease of getting new land.

Whereas various forms of peasant agriculture can support up to one hundred people per square mile, the typical rainforest cattle ranch employs one person per two thousand cattle, which, since the stocking rates are so low, amounts at best to one person per twelve square miles. Meat production barely reaches fifty pounds per acre per year. In northern Europe, on farms that don't use imported feed, meat production is over five hundred pounds per acre per year, plus one thousand gallons of milk. There is no milk production on rainforest ranches.

In most Latin American countries, the simplest way to establish title to unoccupied land is to cut down the trees. In legal terms deforestation is improvement of the land. Peasants do most of the clearing, though most of the land is subsequently occupied by big ranchers. Many of the squatter-clearers deliberately strip the land with the intention of selling it off, albeit at a low price set by the buyer.

It is not easy work. After the backbreaking job of cutting down all but the smallest trees comes the task of setting the tangled mass of wood alight. It may then take several days for the fire to die down —it may take several weeks—and the fire may go out of control. In 1974 Volkswagen do Brasil started a forest fire of 3,900 square miles on its ranch south of Tucurui by spraying it with defoliant and setting it alight. The fire raged for many weeks, but such is the difficulty of patrolling the area that Brazilian officials only learned two years later, from a satellite photograph, that they had been hosts to the biggest forest fire in Brazil's history.

When the transient fertility of the newly burned land wears off, in a year or two, and tenacious weed species creep back from the surrounding forest, a peasant attempting to farm may choose to sell up to a rancher, if he can find a buyer. Otherwise, he may abandon his plot and try to find another area to clear. A rancher can probably continue profitably on land of declining productivity for another three to five years, burning repeatedly to keep down poisonous weeds and encourage new grass to grow. Weed control accounts for one quarter of the operating costs of a rainforest cattle ranch.

A pasture abandoned after repeated burnings is permanently degraded. Even if the land is subsequently left fallow, it is unlikely to produce anything but a covering of rough scrub. The pattern of clearing, repeated burnings, falling productivity, and abandonment is now familiar to ranchers. Nearly all the ranches established in the Amazon before 1978 are now abandoned. But with cheap land, cheap labor, low inputs, and good prices for beef, it is nonetheless profitable to ranch using this system. Among the companies doing so are Heublein, Liquigas, and a consortium that includes Armour and the King Ranch.

"It is still good business to clear virgin forest in order to fatten cattle for say five to eight years and then abandon it," Budowski comments. "By clearing the forest you make a little money from the timber. And you can get people to do it for you by letting them use the land for one or two years to plant corn or what have you before you put in the grass. So it costs you nothing to clear the land, and after a few years you move out to repeat the whole thing somewhere else. This means that there is no incentive for small farmers to be permanent on their land since they know that they will be able, as the land's productivity goes down, to sell it off to the cattle ranchers. In fact it is often the case that they only have the land on loan, so to speak, from a rancher who provided them some advance money to clear it. This system certainly is the case in Brazil. It was the case in Venezuela when I studied it. I know it's the case in Colombia. I suppose it's the case here in Costa Rica, too, although the pattern may vary a lot from one place to another.

"Land degradation caused by grazing is by far the greatest challenge in Latin America," says Budowski. "I look in horror toward the day the tsetse fly is controlled in Africa. What's happening here in Latin America will happen on a much greater scale in Africa if large ranches are established on forestlands there. I would welcome the disappearance of cattle except for dairy farming. Dairy cattle have a place because of the milk, but sometimes I like to think of ourselves as one of the last generations to eat beef. We cannot afford the luxury of growing grass and then converting it into meat at a ratio of 5 to 1 or 6 to 1. We should rather eat the grass ourselves, which means growing other crops instead of grass for cattle."

. . .

Thirty years ago a group of American Quakers from Fairhope, Alabama, came to Costa Rica and founded a dairy farm in a cloud forest. The twenty-five-mile journey north from Puntarenas on the Pacific coast to the Quakers' farm in the Tileran mountain range is easier today than it was when they made their first trip. The original track was so rough that the settlers were forced to turn to cheese making because they could not get their milk to the towns on the coast in a salable condition. The oxcart trail through the mountains has been improved so that in the dry season most four-wheel-drive vehicles have no difficulty negotiating it. In the wet season everyone must take his chances. For those without a jeep or Land-Rover, a decommissioned schoolbus travels the steep, twisting track daily to the nearby settlement of Santa Elena. The bus breaks down on roughly one journey out of every two. And there are other hazards. On my trip back down the bumpy road to Puntarenas, the woman in front of me threw up, mostly on my shoes, for the whole of the three hours it took us to descend the mountain.

The first colonists, forty of them, came on foot and on horseback. Each family bought its own plot of land. They also set aside more than one thousand acres, the headwaters of the Guacimal River, as a communally owned forest reserve so as to protect their water supply. In all the community possessed about three thousand acres. In a small clearing they set up a meeting house and storehouse. Families pitched tents in the mud and began clearing land and planting gardens.

Five years after the first wave of settlers, Arnold and Mildred Hoge arrived from Iowa. "Mildred is related to half of the first families who came here," her husband says. The Hoges seem to have stepped out of a Norman Rockwell painting, he in overalls, workboots, and a hunting cap, and she in an apron, smiling, her gray hair pinned up in back, looking like the first-prize winner in the Betty Crocker Bake-off. They built their house themselves, and it is filled with furniture that Arnold made and Mildred upholstered. On one table is a pile of books and magazines. I notice a copy of the magazine of the Costa Rican Association for Nature Conservation in the pile. Arnold pulls out a book and hands it to me: *Costa Rica—A Country Without an Army* by Leonard Bird. "He's a peace person," says Arnold, referring to the author. He smiles at me and leans forward. "Are you a peace person?"

The Hoges are enthusiastic about their adopted country. "You know, Costa Rica has one of the best literacy rates in the world," Hoge remarks. "Ninety percent. Why do you think that is? Because they spend their money on social programs—medicine and education—instead of on weapons of destruction." In 1948 a well-to-do farmer, José Figueres Ferrer, led a successful revolution after the government annulled a close-run presidential election that appeared to have been won by the opposition party. Figueres revised the constitution to abolish the army and then, Cato-like, after eighteen months as head of government he handed power back to the winner of the 1948 election.

"Oh yes, no army," says Hoge. "That was a big attraction to the original settlers. That's how they first got interested in coming to Costa Rica. A couple who had been to Costa Rica were in Alabama and they addressed the Fairhope meeting. Now some of those people, like Wolf Guindon, had just come out of prison for refusing the draft. After they heard this couple, they got interested in coming out here. They joined up with Mildred's people, from Iowa, and started looking for land they could farm on." On April 19, 1951, one of several search parties the group had formed found what they were looking for—a mountain straddling the Continental Divide, 4,600 feet above sea level, covered by one of the most dense, most lush rainforests in the world. They called it Monteverde and decided to start dairy farming there.

The group's decision to leave home and found a Quaker community in a country most Americans had never heard of attracted a fair amount of publicity in the United States. There was a steady stream of inquiries from people interested in joining them. After they had settled in at Monteverde, the group composed a statement of aims and ideals, which they sent in response to such inquiries. It explained that they had moved to Costa Rica because of "our desire to leave behind the constant worry of war and free ourselves from our increasing involvement in militarism and government control in all phases of our life." Searchers after Utopia were told of the "unbelievably poor road" up to Monteverde and warned that "it is often a year or more between visits to towns or cities." The statement went on to say that "at present farming is done by hand and anyone who holds any feelings that manual labor is inferior or insignificant, and is not willing to identify himself with the working people of the community, should look elsewhere for a location. . . . There should be an attitude of cooperation for the common good of Costa Ricans. This includes conservation rather than exploitation of land, timber, and other natural resources. . . . We hope to find a way that would seek the good of each member of the community and live in a way that will naturally lead to peace in the world rather than war."

In many ways Monteverde today is a relic of Middle America at the turn of the century, when there were still barn raisings, community songfests, and hand-cranked telephones, and when people walked where they wanted to go. Many of the original settlers still speak no Spanish, though most young Monteverdians are bilingual. The spiritual and material hub of the community is a collection of well-designed wooden buildings, the meeting house, the school, and the library. Their plate-glass windows look out onto a grassy field on one side and a shady wood on the other. Monteverde is not a commune, and no one is obliged to take part in group activities, but with only themselves to rely on, the community plays and works together a great deal. One morning while I was there, a pancake breakfast to raise funds to replace a dangerous bridge was followed by a spring cleaning of the school and library. Fifty or more men, women, and children spent the morning washing floors and windows, repairing the school roof, reordering the library, and cleaning out the woodworking shop. There is an elected town meeting with a chairman, but it operates entirely by common consent and has no control over individual land ownership or over who can move into the community. The original founders are mostly past retirement age by now, and a second generation, many members of which were born and bred in Monteverde, is taking over.

Monteverde is still spoken of as a Quaker community, but there is now a sizable contingent of Seventh Day Adventists as well as a number of independent settlers who might be classed under the general heading of aging hippies. Ironically, although Quakerism might seem to have more in common with the flower children mentality than with the fundamentalism and proselytizing of the Adventists, it is the latter, with their Quaker-like firmness of purpose, who have integrated most successfully into the community. "Those others," Arnold Hoge says, shrugging, "are nice people, but they aren't too good for the community. They usually give up when their money runs out and go back to the States, where they can earn a living."

Walking along a dirt track that follows the gentle curves of the mountainside on my way to see the Hoges gives me a glimpse of country life as Jane Austen must have known it. There are very few cars or telephones, so paying visits is neither quick nor foolproof. Almost any errand involves a walk of several miles, with no certainty that the person one is coming to see will be there at the end. There are no distracting noises and it is pointless to hurry. Certainly Elizabeth Bennett looked out on no more pleasant farmland than Monteverde's green meadows and gentle hills, with their peacefully grazing Guernseys and Holsteins.

When the Quakers arrived at Monteverde, most of the land was virgin forest. Here and there squatters had cleared and burned small patches for gardens. Soil tests showed that the black volcanic soil of the region was sandy enough to be workable and was seriously deficient only in phosphate. Monteverde's moist climate was an advantage in creating the lush pastures that dairy cattle require if they are to produce a steady supply of milk. But when they took over the previously cleared patches of forest, the founding families noticed how poor the soil was wherever the forest had been cleared by burning. Burning, they realized, destroyed the humus, the decayed organic material that forms the top layer of the soil and provides the nutrients essential for plant growth. So the first rule at Monteverde was No Burning. Initially all clearing was done only with axes, machetes, and brush hooks, though eventually the bigger trees were felled with chain saws.

The Quakers had respect for the principles of conservation from the beginning. Arnold Hoge, who began farming in 1936 in Iowa, attributes their awareness to lessons learned when they were starting out. "That was at the bottom of the Depression. The Soil Conservation Corps was just starting up. Oh, yes, we learned the value of conservation back then. Our farm was all in contours."

The Monteverde method of creating pasture from forest is straightforward but, in tropical terms, revolutionary. First they clear away the undergrowth and seed selected grasses, which are sheltered from the full strength of the sun by the tall trees. After the grasses have established themselves, the trees are cut down and left to decompose, slowly returning their nutrients to the soil. Some trees are left standing to provide shade and as anchors for the soil in tropical rainstorms. The cows wander among the fallen trees, grazing. The leaves and branches decay first and soon only the trunks and stumps remain. These are broken up with chain saws and left to rot and enrich the soil. The farmers apply phosphates as fertilizer to make up for the lack of phosphorus in the soil. Some have adopted a practice of rotating pastures, which enables them to use less fertilizer and keep more cattle on the same amount of land. The idea is to divide your pasture into thirty equal parts. The cows graze on each area for only one day before being moved on, and each area is fertilized as soon as the cattle leave. In this way there is no problem with overgrazing, since each area is grazed for only twelve days a year and each has a month in which to recuperate and give the fertilizer a chance to work.

The Quaker settlement has had a strong influence on the surrounding Costa Rican farmers. Most of the local farmers have switched from subsistence farming to dairy cattle to supply Monteverde's *lechería,* or cheese factory, which needs 2,300 gallons of milk every day to produce two thousand pounds of cheddar, gouda, and their own cheese, called Monterico. Local families also supply husbands and wives for second-generation Monteverdians, and an entrée into Costa Rican life for the whole settlement.

To a pessimist it may seem that the seeds of Monteverde's decline are sown in the philosophy that has made it successful. The dairy farms and the *lechería* have prospered on a combination of conservation and relatively expensive modern inputs of fertilizer and feed concentrate. But in recent years the price of milk has held constant while the cost of fertilizer and concentrates has tripled. The logical way to bear these increased costs is to expand and farm more intensively. But most Monteverdians subscribe to the small-is-beautiful philosophy. They want to be in control of their lives and their work, want it to be on what they consider a human scale. "We like the community the way it is," one young farmer told me. "We don't want to expand too much or fast. There's no point in saving the community if you have to destroy the ideals it was founded on." Not everybody in Monteverde is a farmer. The Hoge's son is now the accountant for the *lechería.* But as the children of Monteverde grow up and take to farming, there is beginning to be a problem of dividing up the existing land. One solution may be to diversify. A study financed by an independent development agency, the Inter-American Foundation, is looking at the possibilities for other money-making activities in Monteverde and the surrounding region. The suggestions include market gardening, forestry, producing biogas from cow manure, and such craft industries as ceramics and weaving. Observers of the efforts of development agencies may see such a study as the most desperate indication that the end is nigh for Monteverde.

The forest attracts many scientists and what might be termed scientific visitors, naturalists and students, to Monteverde. As the mountain road has been improved, it has become easier to take the cheeses to market, to bring supplies to Monteverde, and to get adequate medical attention for sick people. It has also become easier for scientists, tourists, and would-be settlers to make the trip to Monteverde. By and large, the settlers get on well with the researchers who come to work in the reserve. They are somewhat more worried at the prospect of more and more tourists coming to see the reserve once the road is improved over the next decade. Most people seem to feel, though, that with some planning and controls, Monteverde will be able to handle more tourists in the future. Monteverdians take pride in the interest their environment inspires, although, like residents of most other tourist meccas, they seldom visit their own most famous attraction.

Just above the farms is the community's own nature reserve, where the forest has been left standing. There biologists are studying one of the wildest and most mysterious natural systems left on earth, a tropical cloud forest. George Powell, a biologist and friend of Wolf Guindon, led a campaign to extend the Quakers' original 1,250-acre forest reserve in order to protect the habitat of a species discovered only in 1964. The male golden toad (which is actually a brilliant orange) is the most brightly colored of all toads, and it lives only in the Monteverde cloud forest. Conservationists raised funds to buy another 5,000 acres, and in 1972 the whole reserve, which is now owned and run by the private Tropical Science Center in San José, was given protected status by the government.

People are often disappointed when they see a tropical rainforest for the first time. Many of them do not live up to the popular image of a tangle of lush, dense greenery dripping with moisture. Some are seasonal; that is, they undergo a rainy and a dry season. During the dry season many trees lose their leaves and the sun streams in through the bare branches. To the uninitiated the forest then loses much of its depth and mystery. But a cloud forest, so called because a blanket of low cloud hangs in the trees most days of the year, never disappoints. Cloud forests are very humid and very conducive to growth. In Monteverde every inch of every tree trunk seems to be covered with epiphytes, vines, climbers, moss, or some other form of life that my ignorant eyes failed to distinguish in the jumble of green. Although it has not been exhaustively examined, the Monteverde reserve is known to contain more than 2,000 species of plants, 320 species of birds, and 100 species of mammals. By comparison, the British Isles, every inch of which has been eagerly combed by enthusiastic amateur and professional naturalists for several centuries and which is twelve thousand times the size of the Monteverde reserve, has fewer than 1,500 native species of plants, 214 species of breeding birds, and 65 species of mammals. The most spectacular bird at Monteverde is the beautifully named resplendent quetzal. Crimson-breasted, with iridescent emerald green tail feathers trailing two feet behind it, the quetzal is the bird the Aztecs worshipped as god of the air. Jaguars and ocelots also roam the forest, as do their prey—tapirs, brocket deer, peccaries, and agoutis.

. . .

One hundred thousand years ago all the mountains between Puntarenas and Monteverde were covered with such forests. So they stayed, until about thirty years ago, when people began to move up from the valleys with cattle. The ranchers replaced the forests with pastures, cutting down and setting fire to the trees. As the lower slopes became more crowded and soil fertility declined, people moved farther and farther up into the hills, cutting and burning for pasture as they went. Today the process of deforestation is almost complete. The few miserable, scraggly trees that remain emphasize rather than disguise the extent of the devastation. They are not the proud specimens that stand guard over the playing fields and pastures of the English countryside. The remnants of forest here are all too clearly leftovers, forest trees intended to be one of a crowd, surrounded by countless other trees. On their own they seem naked, shriveled, lost. Nor is the ground cloaked with the lush green grass of an English meadow. The land is brown and dry, covered with scrub in some places, completely bare in others. Arnold Hoge has a one-hundred-acre wood on his property, in which 120 species of plants grow wild. Bill Haber, a botanist from the University of California who has botanized all over Costa Rica, told Hoge that, apart from official nature reserves, his hundred-acre wood is now the largest area of uncut forest left on Costa Rica's Pacific slope.

Budowski thinks that some of this denuded land, though not the very steep hills, could be farmed productively over a long period if people would change the habits of centuries. "The answer for this area is not cattle ranching or conventional high-input farming. We believe now, but I can only say *believe,* that the answer has to do with the intensification of agriculture based on a very careful management of the land, which includes, for instance, using plants that fix nitrogen, and a judicious combination of crops instead of planting in monocultures, and integrating crops and trees to increase production."

The biggest obstacle to increasing agricultural productivity is not technical but social. Researchers may conclude that it makes sense, for example, to plant a certain crop that enriches the soil and produces high-protein fruits. But farmers are a conservative breed, and the poorer the farmer, the less willing he is likely to be to risk his precarious livelihood by experimentation. The new crop may have performed well in test plots, but the farmer will want to know how it will do under the less than ideal circumstances that prevail on his and his neighbors' farms. If he has a surplus, will he be able to sell it? Does its vigorous growth depend on expensive fertilizers or time-consuming management? Once it has been harvested, is it easy to store? Does it taste good?

Another difficulty, according to John Palmer, a British forester at CATIE who has also worked in tropical Africa and Asia, stems from the imposition of European land-use systems on Latin American soils. "There is virtually no local tradition of good land use here. The indigenous communities have been either bypassed or wiped out. Very few of their traditions have been integrated into the Iberian agricultural experience. In Southeast Asia, by contrast, the experiences of the local people over hundreds and thousands of years still have an impact on the way of farming."

The British Empire was, it may be argued, wider than it was deep, an overlay on the cultures it ruled. Spain sought cultural and ideological dominance over its empire, but Britain's interest in its colonies was largely mercantile. Today, for example, Roman Catholicism is the dominant religion of former Spanish colonies, whereas local religions still hold sway in India, Malaysia, Kenya, and other ex-British colonies. In San José and Brasília, and throughout the Latin tropics, dentists and lawyers establish their manliness and their connection to the aristocratic Spanish past by becoming cattlemen. Cattle ranching makes no sense for large parts of the continent; nonetheless great areas of land are being degraded, and millions of peasants are short of food because of the Iberians' overwhelming atavistic urge to become a *caballero.*

Palmer doesn't believe it can last. "Allow me to be philosophical. I don't think that people can continue in the more highly populated areas of the tropics—maybe I should say of the Latin tropics—to hold the same philosophy of land use which their forefathers did when life expectancy was much lower and child mortality was much higher. There will have to be a shift to a more Southeast Asian view of life. There people have been living in crowded communities for hundreds of years. The experience of living in close proximity has forced people into communal land use, and they have evolved very efficient techniques to cope with high population densities.

"It does seem unreasonable to me now that everyone should expect to have the right to own a plot of land that will enable him to live completely independently of his neighbors. With the rise of population, it is simply not sensible for everyone to have thirty-five acres because inevitably it means moving onto sites which are not suitable for permanent agriculture." ("In Southeast Asia," Budowski had told me, "less than a quarter of an acre is enough for a whole family to live on because they farm so intensively. Here we call a small property something with twelve acres.")

"You can have a stable agriculture on these soils and one which enables you to grow enough extra to get cash for other goods," Palmer went on, "but you probably need to use a much higher degree of management. You need to work the land much more intensively than people here are prepared to do." He shook his head and sighed. "It's no use saying, 'Oh, the land can't stand it,' because the Chinese and other people in Southeast Asia and southern Asia have been doing this for thousands of years. It's the local attitude and local institutions that prevent it. This isn't a technical matter at all. The overriding factor is the ghastly land tenure, which at the moment benefits the big people, and is based on cutting down forests." Palmer, a quiet, infallibly polite, middle-class Englishman, becomes uncharacteristically heated when he speaks of "vicious landowners" who raise rents whenever their tenant farmers manage to improve their productivity. "The Chinese changed their land system, but it took a revolution to do it."

7

The Harvest

For six hundred years, while Western Europe groped through the Dark Ages, a magnificent culture flourished in the rainforests of Central America. The Maya were the only people in the Americas to develop an original system of writing. Their mathematics was many centuries ahead of the European system. The Mayan calendar was more accurate than the Gregorian calendar we use today. Mayan ceremonial buildings, their frescoes, sculptures, and bas-reliefs are still admired for their grace and harmony. At its height, in the eighth century, the Mayan civilization supported fourteen million people. All this took place in the fragile and difficult rainforest environment. The key to it was the Maya's sophisticated system of agriculture.

The heyday of Mayan culture ended around the tenth century. The great stone cities of Palenque, Tikal, and Piedras Negras, apparently made redundant by changes in trade patterns, were then abandoned to the creeping vegetation of the forest that surrounded them. Their inhabitants dispersed throughout the forest, where they and their descendants farmed, fished, and hunted until the coming of the Spanish conquistadors in the sixteenth century. For two hundred years the Spaniards raided Mayan forest settlements to get slaves for their ranches and plantations. Much of the Mayan system of knowledge was lost. The Spanish deliberately destroyed almost all their writings and suppressed their religious practices. The skills and knowledge by which they managed the forest fell into desuetude.

Most Maya of the Chiapas rainforest in southern Mexico were captured by slave traders, and their lands settled by Mayas fleeing north from persecution in Guatemala. Ignored by the Spanish rulers, the refugees, known as the Lacandon Maya, were able to practice a simplified form of the highly productive Mayan agriculture in peace. Thanks only to their guarding the knowledge of their ancestors, we know today something of how the rainforest was able to support a complex civilization and a population of many millions for one thousand years. The Lacandones are gentle people who wear their hair long and dress, both men and women, in plain white cotton homespun tunics. Using ancient Mayan techniques, they farm more successfully and more efficiently than any of the businessmen, colonists, or agricultural experts who have also tried to farm in the forest.

At the beginning of the year, the Lacandon clear small plots from the forest, two to three acres each. They leave the felled trees and cut vegetation to dry. In deciding where to clear a plot, the Lacandon distinguish between seven types of soil, only three of which they consider good for farming. They also take into account the vegetation of the area. Mahogany and tropical cedar are to be avoided because the soils that support them are too wet for agriculture, whereas ramon and ceiba trees grow on rich, well-drained soils. In April they clear firebreaks around each plot and set it ablaze. Since in a tropical rainforest most of the nutrients are stored in the living vegetation, burning the forest creates a nutrient-rich ash in which the Lacandon's crops flourish. But with the forest cover gone there is a danger of erosion. To prevent this the Lacandon at once plant fast-growing trees, such as banana and papaya, to provide shade, and root crops like taro and chayote, to anchor the soil. A few weeks later they plant their staple crop, maize. Along with the maize they cultivate as many as eighty other crops. These include onions, garlic, pineapples, custard apples, chili peppers, watermelons, limes, lemons, grapefruits, oranges, coriander, squashes, lemongrass, yams, fennel, cotton, sweet potatoes, tomatoes, manioc, mint, tobacco, rice, avocados, parsley, beans, plums, guavas, sugarcane, cacao, and ginger, plus many plants unfamiliar outside the tropics.

In order to decide when to plant certain crops, Lacandon farmers watch certain "indicator" plants in the nearby forest. When, for example, the wild tamarind tree flowers, farmers in the northern part of the Chiapas rainforest know that it is time to plant their tobacco plants. This ingenious system, which depends on the existence of the natural forest, works better than a fixed planting schedule, since it automatically corrects for annual fluctuations in rainfall and temperature.

A Lacandon farm plot (or *milpas*) is successful partly because it mimics the forest in its density and diversity. The Lacandon do not plant their crops in tidy, straight rows, separated by category. On the contrary, they make a point of not planting a clump of one plant within ten feet of another of the same variety. The effect is to make the best use of the available nutrients and to prevent the spread of plant-specific pests and diseases.

The spaces in between crops are not kept clear, because there are no spaces between crops. Every inch of a Lacandon *milpas* is covered with growth, several layers of it. Papaya trees shade maize, tobacco, rice, sugarcane, and other medium-height crops. The ground is carpeted with vines, and underneath, in distinct subterranean layers, are the root crops: sweet potatoes, manioc, and yams. The arrangement is not random. Each crop has soil, water, and light requirements, as well as different responses to associations with every other plant. The Lacandon have learned through centuries of trial and error which of the thousands of possible permutations are best on a given plot.

Productivity is high on a Lacandon *milpas.* Measurements by anthropologists James Nations and Ronald Nigh showed that a 2.5-acre plot can yield up to six metric tons of shelled corn each year, after losses to animal pests, and the same amount of root and vegetable crops. The average annual yield is somewhat lower, according to Nations and Nigh, but it is still significantly more than the amount harvested by recent immigrants to the area, including the descendants of the original Maya of the Chiapas forest, the ones who were enslaved by the Spanish. The government is now encouraging these unfortunate people to return to the lands of their ancestors, but centuries of exile and deacculturation have made them strangers in the forest.

The Lacandon cultivate the same plot for from three to seven years, according to how free of weeds they keep it. As long as weeds are kept out, the Lacandon can get two crops of maize a year without significant drops in yield. Lacandon farmers reduce weed invasion by making sure that their plots are surrounded by mature forest, which is naturally resistant to weeds. They also burn plant debris after each maize harvest, a practice that kills weeds and recycles unharvested nutrients. But inevitably weeds encroach on the *milpas,* and sooner or later it is easier to clear a new plot than to weed the old one. At that point the family starts again clearing a new *milpas.*

They do not, however, entirely abandon their old plot. They plant rubber, citrus, cacao, and other trees. During the five to twenty years that the plot is left fallow, the Lacandon harvest such wild plants as balsa wood, old crop plants, and the useful species they have planted there for food, fiber, and construction materials. Perhaps more important, in this stage of its cycle the "orchard-garden" attracts wild animals that the Lacandon depend upon for protein. They are, in a sense, the Lacandon's game reserves. Scientists have found that certain animals, especially deer, squirrels, pacas, peccaries, and agoutis, are more common in regenerating farm plots than in pristine forest. They are apparently attracted by the young vegetation and the planted food crops, although most of them also need the mature forest for certain critical times of their life cycle. The Lacandon also hunt and collect wild plants from the surrounding undisturbed forest. By the time they are ten years old, Lacandon children can distinguish hundreds of edible and otherwise useful plants from those that are harmful.

The Lacandon prefer to clear their *milpas* from regrown plots because it is much easier than clearing mature forest. It takes forty man-days of hard work to clear 2.5 acres of virgin forest, whereas clearing the same area of secondary growth takes only eight man-days. By restricting their cycle of active and fallow *milpas* mostly to secondary growth forest, the Lacandon maintain a productive agriculture without encroaching inexorably farther into the forest.

But the system is under pressure as never before, and it is beginning to break down. Only a few scientists and anthropologists have studied the Lacandon system, and no one except the Lacandon themselves knows how to practice it. Disease and maltreatment have reduced the Lacandon to fewer than four hundred people. And only 20 percent of them still practice their traditional agriculture. The rest have been herded into government-controlled "villages" so that the

Chiapas rainforest can be colonized by poor peasants who do not know how to farm in the forest and by cattle ranchers who do not care.

The breakdown began in the 1940s, when the Mexican government's agrarian reform law put an end to the isolation the Lacandon had enjoyed. The law reclassified the Chiapas lowlands as national territory. Landless people were encouraged to find land there, a policy intended to reduce pressure for land redistribution in the more fertile agricultural areas of Mexico. The Lacandon's claims to the land they had lived on, farmed, and hunted on for centuries were ignored. More than eighty thousand peasant farmers migrated into the forest. Tragically, the peasant farmers' methods are unsuited to the fragile and complex rainforest environment. Some are descendants of the original Maya, but unlike the Lacandon they have not retained the farming techniques their ancestors had developed. They achieve disappointing yields with the monoculture agriculture encouraged by the government and destroy the forest in the process. In addition, many businessmen and prosperous farmers have started cattle ranches in the Chiapas region, taking over and consolidating abandoned farm plots. Because they lack the skills of the Lacandon, clearing the forest and then selling it to ranchers is, although a violation of the spirit of the agrarian reform laws that were intended to give poor people farmland, the only way many peasants can make a living from the difficult rainforest soils. Already the northern third of the five-thousand-square-mile Lacandon forest has been burned and converted to vast cattle ranches. The Lacandon forest itself is one of the only three large lowland rainforests left in all of Central America.

Under pressure from waves of immigrants, the Lacandon moved deeper and deeper into the interior of the forest. But in 1971 the Mexican government forced them to move into three small reservations, with a total area smaller than twenty thousand acres, where it is difficult for them to continue their self-reliant way of life. Now that the Lacandon live in concentrated settlements far from their gardens, instead of each family living in its own *milpas,* they can no longer keep the *milpas* clear by daily weeding. As a result, the *milpas* becomes overgrown much more quickly than before, and the Lacandon are forced to abandon them for new plots much earlier. This is exacerbated by the Lacandon's increased use of machetes. Previously they weeded by hand, uprooting the plants. Now, as they swing machetes through the weeds, the seeds are scattered widely across the ground and the weeds become ever more strongly entrenched. Missionaries and government officials urge the Lacandon to "improve" their standard of living by depending more and more on such manufactured goods as machetes and chain saws. This not only undermines their traditional, independent way of life but also encourages the Lacandon to become low-paid laborers for the ranchers and loggers whose destructive practices generate a cash income.

The Lacandon system is on its last legs, but it is not too late to save it and improve it. No modern agronomist has a better understanding of the workings of the forest than these farmers, but there is still room for more efficiency. The Lacandon do not possess all the skills of the early Maya. According to historians and anthropologists, the Mayan civilization depended heavily on now lost techniques of irrigation, terracing, and intensive aquaculture, appropriate versions of which could be reintroduced. Lacandon farmers would also benefit from the cooperation of modern crop breeders, experts on agricultural pests, specialists in irrigation and water control, and marketing advisors.

. . .

Shifting cultivation is based on field, rather than crop, rotation. It is appropriate where the soil is poor and there is adequate land. Throughout the world where these conditions have prevailed, farmers have practiced shifting agriculture. In Amazonia and Southeast Asia shifting agriculture is at least two thousand years old. Well into the twentieth century, parts of Europe and North America were farmed by shifting cultivation. In England until the Second World War, for example, many farmers in East Anglia, the Lake District, the South Downs, and the Chilterns rotated their fields, cultivating each for two to three years before leaving them to return to woodland for up to twenty years before cultivating them once again. This system ended when land prices rose so high that farmers could not afford to leave any of their land uncultivated, and the availability of inorganic fertilizers at government-subsidized prices made continual cultivation possible. This system too has problems. On many intensive farms now greater and greater amounts of fertilizers are needed simply to maintain existing levels of productivity.

Shifting cultivation is often more efficient than permanent field agriculture. That is, it takes less labor to produce a specific unit of food by shifting cultivation than by other methods of farming. For this reason some civilizations have returned to shifting cultivation when the pressures that forced them to adopt settled agriculture have eased. After the Maya dispersed throughout the rainforest in the tenth century, the Lacandon retained those elements of intensive Mayan agriculture that enabled them to feed themselves most efficiently. They dropped some of the more intensive agricultural practices, such as fish farming and field irrigation, that were unnecessary —and too expensive in terms of labor and resources—for a small population with access to plenty of land—and "free" food in the form of the wild plants and animals that can be gathered and hunted there.

Purely as a hunting and fishing ground, the rainforest can support from five to seven people per square mile. Shifting cultivation increases the overall carrying capacity of the forest to from ten to one hundred people per square mile. Ignoring the large areas that are in fallow at any time and taking only the cultivated areas into account, the figure rises to from two hundred to six hundred or even more people per square mile. Intensive modern agriculture, where it can be practiced, can support higher densities of people, but the productivity of shifting cultivation compares well with other forms of agriculture that do not rely on expensive machines and artificial fertilizers.

The more "primitive" a farming system is, the more knowledgeable and skillful the farmer must be. Shifting cultivators and those farmers who have permanent "gardens" on forest soils traditionally create in their fields a partial replica of the complexity of the forest, its diversity, its several layers, its mixture of plants and animals. But the opposite is true of "modern" agriculture. There uniformity is the key. Vast fields are devoted to a single crop, often indeed to a single variety of one crop. The modern farmer possesses powerful tools—year-round irrigation, machinery, ample supplies of power, insecticides, herbicides, and fertilizers—that have a homogenizing effect on the land. He relies not upon his skills but upon his tools.

. . .

Ed Price, an agricultural economist who works for the International Rice Research Institute in the Philippines, spent three years with small farmers in a village called Cale, in Batangas Province. "Cale was not in the really steep and difficult highlands; it was more rolling hills, but there was no irrigation. In three years we identified more than 160 different crops and crop combinations grown by those farmers. They performed more than one hundred different technical operations. Whereas if you talk to a year-round irrigated rice farmer, he grows probably only one or two varieties of rice. And you can certainly number on two hands every operation he does to those crops. It's very simple, cut and dried.

"But then, if you leap to tribal peoples in the hills, the agriculture is even more complex. They know the names of far more plants, and they grow far more plants. Their pest control strategies are more complex and their planting and harvesting timetables more finely tuned. They are more aware of wildlife in general. Forest species are one of their resources. In fact that is one of the distinguishing features of tribal people, that they depend partly on the forest itself for what they need. They forage, collect resin, gums, rattan, in addition to growing rice and other crops." The Hanunoo people of the Philippines are hunter-gatherers who divide the plants in their territory into 1,600 categories, although botanists can only distinguish 1,200 species.

Successful systems of shifting agriculture and permanent home gardens are the result of many lifetimes of experience, trial, and error. Each generation carefully teaches the next the accumulated skills and knowledge of centuries. But the newcomers who today are flooding the forest don't bring with them this knowledge. They treat the forest as if it were the grasslands with which they are familiar. Their agricultural advisors often go one worse by encouraging them to treat tropical soils like the temperate soils on which *their* training has been founded.

Governments often aid the movement of settlers into rainforests, by building roads into the forest and by offering grants and technical assistance. The aim is usually a dual one: to deflect pressure for land reform elsewhere in the country by opening a new frontier, and to extend the dominant culture into regions populated by groups with their own way of life. Almost by definition the settlers are ill equipped for the difficulties and complexities of forest farming. There are three victims of their inexperience and the bad advice they receive: the settlers themselves, the forest, and the indigenous people (whose knowledge might have saved the first two).

The record of colonization in rainforests is a dismal one. The World Bank has supported many rainforest colonization projects and continues to do so, but even it admits that "successful examples of colonization by nationals [that is, non-forest-dwelling groups] are exceedingly rare." Colonists at Angamos in the Peruvian rainforest, in a project financed by the Peruvian and Swiss governments, are unable to grow enough food to feed themselves. During the regular periods of famine, the colony survives, according to a World Bank document, only by buying, but more usually stealing, food from the Matses, indigenous forest dwellers who are their neighbors. The Matses are looked on by the colonists and project officials as primitive people. They have only simple tools, whereas the colonists have more sophisticated tools as well as fuel oil, pesticides, and fertilizers. Nonetheless, the Matses grow enough food to feed themselves and survive the predations of their subsidized neighbors. In spite of this and similar incidents elsewhere, colonists are never encouraged to adopt the agricultural methods, or even modifications of them, of the local people.

Settlers imported into the Brazilian Amazon in the early 1970s on a government colonization scheme associated with the newly built Trans-Amazon Highway invariably chose to establish their farms where the trees were thick-trunked and gave the impression of vigorous growth. In fact these areas were always associated with clay-poor red earths, the typical nutrient-poor soils of the rainforest. Traditional shifting cultivators in the area, however, looked for certain thin-trunked species that, scientists later discovered, indicated well-drained, clay-rich soils. The local shifting cultivators also grew a different mixture of crops. After one year of farming, the soil of the native farmers proved to have superior chemical composition to that of the newcomers in every respect. The native farm income was twice that of the settlers, who had had the benefit of government agricultural advisors.

Immigrants to the forest, like the Brazilian settlers just described, are practicing a distorted kind of shifting cultivation that is called slash-and-burn agriculture to distinguish it from the balanced shifting cultivation of the true forest dwellers. Most of the two hundred million or so people farming in or on the fringes of rainforest are slash-and-burners who are used to conventional agriculture and don't have the skills and knowledge of the indigenous forest farmer. About one half of the primary rainforest that is destroyed each year is ruined by slash-and-burn farming.

The worst loss is in Southeast Asia, where slash-and-burn farmers now clear thirty-three thousand square miles each year, although much of it has been logged or cleared before. Almost two thirds of the twelve million square miles of land thus farmed is in the hills, where forest cover is most crucial. Africa's slash-and-burn farmers are estimated to clear fifteen thousand square miles of rainforest every year. Ghana, for example, has virtually no rainforest left. At one time more than one third of the country, thirty-five thousand square miles, was rainforest. Today, apart from about six thousand square miles of ill-protected forest reserves, less than two hundred square miles remains, and slash-and-burn is the main cause of the destruction. In South America, large government-subsidized cattle ranches, not landless colonizers, are the main cause of deforestation, but colonization schemes have hit certain areas hard. Worldwide, according to a 1980 National Academy of Sciences study, slash-and-burn destroys forty thousand square miles of primary rainforest each year.

The lost diversity of traditional subsistence agriculture was remembered by a Sri Lankan farmer, Mudiyanse Tenakoon, who spoke in 1982 with the British ecologist Edward Goldsmith.

> I can remember 123 varieties of red rice; now only three or four remain. First of all we needed different varieties for the two growing seasons—the Maha season associated with the North-East monsoon and the Yala season associated with the South-West monsoon.
>
> During the Maha season we planted what we call the "four month" varieties. As their name indicates, they take four months to grow. During the Yala season we planted "three month" varieties. . . . Among the "three month" varieties I can remember Heenati, Dahanala, Kokkali,

Kanni murunga, Pachha perumal, Kuru wee, and Suvandel. We also grew Mawee, a "six to eight month" variety. This was for the priests. Buddhist priests don't eat after noon, so they need very nutritious food to sustain them until the next morning. Mawee is very nutritious; it has a high protein content and that is why we grew it.

We grew Heenati for lactating mothers as it makes them produce more milk and also better milk with a high fat and sugar content. We tried to grow it during both seasons. Kanni murunga we grew for the men going out to work in the paddy fields. It gave them energy as it contained a lot of carbohydrates. It was also used for making milk rice for traditional ceremonies. Suvandel we grew because of its extraordinary fragrance.

Some of these varieties were specially used when there was a lot of water in the paddy fields; others when there was little water. . . . Some varieties were grown when the fields were particularly muddy; some were more suitable to grow on high ground where there was less mud. Some of the varieties required very rich soil; others would do well in the poorest of soils. Some were more resistant than others to the paddy bug and we planted them, rather than the other more desirable varieties, when the traditional means of controlling the bugs had failed.

There is another advantage to the old system; it is that we used to produce all sorts of foods that we cannot produce anymore. To begin with we used to go into the jungle to get many foods such as the Baulu, Weera, Jak fruit, Himbutu, Wood Apple, Wild Pear, and Avocado. Now the jungle has been cut down, we no longer have access to those foods. We must try to recreate the jungle.

We also used to obtain a vast variety of fish from the streams, the tanks and the paddy fields when they were flooded. Some of these fish such as the Lula, Kawaiya, the Hadaya, and Ara, could live in dried up ponds. In this area at least they have nearly all disappeared, some of them eaten by the Tillapia [fish] that have been brought here from Africa and foisted upon us by the government. The government insists that Tillapia eats only vegetable matter, but this is not true. Others, especially those that live in the paddy fields, have been poisoned by pesticides. Since there are no longer any fish, the larvae of the mosquitoes that transmit malaria can now survive the dry period. As a result malaria has become a lot more serious problem than it was.

The Lula that used to thrive in the tanks was also of great value to us because it favoured the formation of blood. That is why we always fed it to pregnant mothers. There were other fish that we obtained from the tank, the Korale, the Petiya, the Hirikanaya, the Walaya, the Anda, and the Ankutta. The Korale in particular was a very sweet fish. Now we only have the Tillapia; it is not bad, but it does not replace all the traditional species, all of which had special uses. . . . The change has unquestionably impoverished our diet and also our lives.

. . .

Ironically, the simplification of modern agriculture, its substitution of uniformity for diversity, makes things harder, not easier, for the small farmer and increases hunger and unemployment. Because modern farms are easier to operate, one person can control more land with fewer workers. "In the northeast of Brazil it is the predominance of latifundia [large holdings], which can employ only so many people, that makes it difficult for people to get their own land and contributes to a movement to other parts of the country, including Amazonia," Herbert Schubart of Brazil's Institute for Amazonian Research, told me. The same is true of southern Brazil, the country's other main farming region. Much of the migration to Rondonia, one of the fastest-growing parts of Amazonia, is due to the big southern farms switching from growing coffee, which is labor-intensive, to growing soybeans, which is highly mechanized.

But though large modern farms are more streamlined than traditional ones, they are less productive, in spite of the fact that small farmers cannot as easily afford yield-increasing inputs. According to the World Bank, small farms in Argentina, Brazil, Chile, Colombia, Ecuador, and Guatemala produce from three to fourteen times more per acre than do large farms. A 1979 study by Albert Berry and

William Cline, using detailed data about the productivity of farms of various sizes, concluded that simply dividing the available land equally among farmers would result in an increase in agricultural output for Malaysia of 28 percent, for the Philippines of 23 percent, for Colombia of 28 percent, and for the Brazilian northeast of a staggering 80 percent. And because the aim of a large farmer is not to grow his own food but, in a sense, to grow money, big farms tend to produce luxury items for export to the people and countries with extra money to spend. Such products are more profitable, though not as essential as basic food crops. With more and more of the best land devoted to export crops, many developing countries cannot feed themselves. Brazil is a good example. She is, after the United States, the biggest exporter of food in the world. But, while Brazil's big growers sell hearts of palm, mangoes, and papaya to Parisians and New Yorkers, 38 percent of Brazilians are malnourished and the country must import a growing proportion of its basic foods. The situation would be worse if Brazil's small farmers were as wasteful of resources as their more powerful colleagues.

Agricultural research—even in the tropics, where millions are subsistence farmers—is geared to the large-scale, energy-intensive cash-crop farm. Rubber is Malaysia's most important export (earning nearly $4.5 billion in 1979), and the country has a Rubber Research Institute established specifically to help rubber growers. Sixty percent of Malaysia's natural rubber crop comes from farms smaller than ten acres. But until very recently all the institute's research work was oriented to big plantations.

Agriculture, and agricultural research, are more subsidized in the industrial nations than they are in those countries whose people depend most heavily on farming. Few have allocated more than 10 percent of their national budgets to agriculture in recent years. It is not the case that farmers in the temperate countries are more enterprising and open to improvements than are subsistence farmers. In the United States, for example, where the federal government has poured $15 billion into an extensive soil conservation program over almost half a century, only 25 percent of the nation's farmlands are adequately protected against erosion.

The Lawa in northern Thailand are shifting cultivators who live in permanently settled villages and have farmed the same land for centuries. They grow more than eighty food crops, plus another fifty for medicine and ceremonial and household uses. In addition, they collect and use more than two hundred wild plants that grow in their fallow fields. Their system supports about 80 people per square mile, taking fallow land into account. One square mile of cultivated land supports 625 people, a ratio that compares well with, for example, Britain, which has one square mile of agricultural land in use for every 750 people. Britain, of course, imports 60 percent of the fresh fruit, 20 percent of the grain, and 23 percent of the meat its people consume, whereas the Lawa are self-sufficient in food.

The Lawa take great care to conserve the land they farm by, for example, carefully controlling all fires, minimizing soil disturbance while working in the fields, terracing, and taking other measures to reduce soil erosion. According to Terry Grandstaff, an advisor to the U.S. Agency for International Development, they possess "incredible amounts of information concerning their environment." Even young Lawa children, says the anthropologist Peter Kunstadter, "distinguish successfully between the 84 cultivated varieties plus 16 useful uncultivated varieties, even at the stage when the plants are less than a centimeter in size."

The Lawa's self-image and view of the world are influenced greatly by their attachment to shifting cultivation and to the land they have farmed for so long. But they are not inflexible. In recent years many of them have adapted to changed conditions and new opportunities by planting new crops and improved varieties, by irrigating their land where possible, and by working for wages and growing cash crops. Some have even altered their religious beliefs so that fewer animal sacrifices are needed.

Increasingly lowlanders are moving into the areas the Lawa have long farmed. And as they become involved in the dominant cash economy, the Lawa's way of life is changing faster than their trial-and-error methods of agricultural innovation can keep up. The Lawa and traditional farmers in general need help from agricultural researchers if they are to adapt in time. Science has a lot to offer them, but so far little has been done. Dennis Greenland, deputy director of the International Rice Research Institute, one of the homes of the so-called green revolution, argues that "the full potential of mixed and relay cropping has yet to be realized. Considerable advantages are to be expected when breeding programs produce plants adapted to production in the mixed crop situation." Scientists, argues Greenland, could help shifting agriculturalists and slash-and-burn farmers by breeding crops that resist certain pests and diseases and that capture nitrogen from the atmosphere to make up for what is lost from the soil. These and other improvements, such as the development of deeply rooted crops that can gather nutrients from far below the topsoil, could, Greenland calculates, increase the productivity of the poorest farmers by up to 1,600 percent.

The most important thing now is to enable slash-and-burn farmers and those shifting cultivators whose system is breaking down to have a stable, long-term agriculture on forest soils. This would mean the disappearance of still more forest, but at least the deforested areas could be productive, instead of being, as now, simply degraded beyond redemption. Even so, though stable rainforest agriculture could take pressure off other parts of the forest, it will not protect the forest from exploitation by settlers or businessmen.

There are a number of things that could be done to help slash-and-burn farmers. They need crop varieties that give higher yields and contain more protein. They need to understand better what nutrients their soils lack and how and when best to add them. They need well-designed and cheap tools. They need better ways of protecting their harvests against pests. They need advice and help in establishing markets for their produce. They need better ways of controlling soil erosion. The list could go on. They would benefit from knowing how farmers in similar environments elsewhere manage their farms. They would benefit if scientists understood the principles that make mixed-cropping schemes so effective and explored ways of improving them.

Many scientists are discouraged by the "hurly-burly" of tropical agriculture as compared with the relative order of temperate agriculture. Dealing with a simpler environment, and one that is made more uniform by the use of fertilizers and pesticides, gives researchers the satisfaction of knowing that their work will be widely applicable. This is less likely to be true with traditional tropical agriculture, one of whose strengths is its variety—its adaptation to local ecological and cultural conditions. This makes fieldwork all the more important in tropical areas, where unfortunately climate and logistical problems make it particularly difficult. For most rainforest regions, the basic soil, wildlife, weather, and sociological studies, on which agricultural research should be based, have still to be done. Some researchers are daunted by the fact that science alone will not solve the problems of traditional agriculture: Scientists will have to give up some of their sovereignty and work with anthropologists or sociologists to avoid the familiar pitfall of promoting technical solutions that for cultural reasons are ignored or have unintended effects. All these factors, along with the tendency of governments, big businesses, and even international aid agencies to concentrate their resources in the cash economy, have combined to leave research on traditional subsistence agriculture on the sidelines.

Nonetheless some progress has been made. At Yurimaguas, in Peru, a team of Peruvian and American researchers headed by Pedro Sanchez in the Amazon Basin have experimented with continuous cultivation of rice, soybeans, peanuts, and maize in an area where shifting agriculture is the traditional method of farming. In eight years the researchers produced twenty-one harvests in each of several fields. When the fields were not fertilized, yields dropped to zero after the third consecutive crop, but with "complete fertilization"—that is, when lime and fertilizer were applied at about the same level that is required to grow these crops on similar soils in the southeastern United States—yields stayed up. Several shifting cultivators participated in the experiments, following the researchers' system on their fields. They obtained similar yields, six to ten times higher than their usual average of 0.4 ton per acre per year.

The trouble with this system, besides the fact that it depends on expensive chemicals that may not always be available, is that it ignores the cost of building 1,200 miles of roads across the Andes in order to make fertilizer deliveries possible, and the cost to the government of subsidizing the fertilizer. Even though every dollar borrowed to buy fertilizers increased a farmer's profits by from $1.29 to $4.95, according to the calculations of the research team, few of the local farmers could afford to borrow money in the first place. Most of those who adopted the new methods for the experiment have gone back to their original way of farming. To address these problems, the researchers are now experimenting with low-cost natural fertilizers and other ways of reducing the need for fertilizers. Growing crop varieties whose natural tolerance of acid soils has been increased by selective breeding "is expected to substantially decrease" the amount of lime and phosphorus fertilizer needed. Rice and cowpeas are yielding good results in this respect. Another low-cost technique being tried is the use of the fast-growing, nitrogen-rich plant kudzu as a natural fertilizer. This "has often resulted in crop yields similar to those following complete fertilization," though growing, transporting, and applying the kudzu requires more labor than applying inorganic fertilizer. Other promising experiments include fertilizing the soil with compost made from crop residues and planting kudzu on worn-out cropland as a quick way of restoring its fertility. It appears that two years of kudzu has the same restorative effect on old cropland as twenty-five years of forest fallow.

Among the other useful discoveries the Yurimaguas team has made are that the soil quality actually improves under continuous cropping with fertilization. After eight years the soil was less acid, leading to a doubling of the soil's ability to retain artificial fertilizers. Levels of toxic aluminum decreased and levels of other nutrients—including calcium, magnesium, phosphorus, and zinc—rose. The fields used were cleared from a seventeen-year-old secondary forest, and trials showed that clearing the vegetation manually and then burning it is better than removing the trees with machines. The burning returned the forest nutrients to the soil, whereas bulldozers compacted the soil and often pushed the topsoil off the field.

The researchers tried growing one crop continuously in the same field but found that yields fell due to a buildup of pests and diseases. These were not a problem when three crops were rotated, although the researchers warn that even rotated crops are likely to develop more pest problems as time goes on. A rotation of rice, peanuts, and soybeans was the most successful regimen, yielding an average 3.3 tons of grain per acre per year. The experiments were conducted on nearly level land, thus avoiding serious problems with erosion and runoff. Nonetheless, the scientists warn that even on flat terrain, crop fields should be surrounded by forest reserves in a mosaic pattern in order to protect the watershed.

Under a United Nations research program, scientists have reintroduced an ancient Mexican system of farming still practiced in the valley of Mexico to four rainforest areas where shifting cultivation was breaking down into slash-and-burn agriculture. The system, called *chinampa,* is based on raised fields built up from soil containing organic debris, usually dredged up from lakes or swamps. The plots, or *chinampas,* are separated by a network of artificial water channels that are used for transportation, fishing, and irrigation. Trees line the channels and keep the *chinampas* in place. The soil of the *chinampas* is regularly replenished by nutrient-rich dredgings from the channels. The *chinamperos* grow a wide variety of crops, and they also hunt and gather plants in the adjacent rainforest. They plant seeds in special seedbeds. When seedlings are ready for the main fields, they are planted out in the little cubes of mud in which they were grown, thus taking their own fertilizer supply with them.

After establishing the *chinampas* and finding that the system worked in other parts of the country, the scientists in 1976 handed them over to local farmers, who have since constructed more. As a low-capital, labor-intensive, self-sufficient system, *chinampa* could have applications in other countries and on other continents.

A number of other small-scale field experiments are under way, aimed at enabling slash-and-burn farmers to stay in one place permanently. In Papua New Guinea, researchers at the Wau Ecology Institute are working with local women, who do most of the farming, in an effort to find a way of fertilizing fields with readily available materials. Most Papua New Guineans keep pigs, but since they are free ranging, it is not feasible to collect their waste for manure, nor do families produce enough kitchen wastes to fertilize their fields. Instead, the institute system relies on waste from coffee plantations, common in Papua New Guinea and usually just dumped into a river. Using the composted coffee wastes, the women build up rows of mounds that follow the contour of the land. The compost-filled mounds provide nutrients to the crops, usually a staple root crop grown in combination with an early crop to shade out weeds and a legume plant to add nitrogen to the soil. The contour mounds also give protection against soil erosion. This is just one aspect of a system that might also include trees grown to reduce erosion, provide food, fuel, and other useful materials, and add organic matter to the soil by shedding leaves and twigs. But there is a cultural hitch to the system. In New Guinea whereas women do the planting, weeding, and harvesting, it is the men who make new garden plots. So, although the system makes sense in theory, it may fail in practice because it requires a change in the balance of work between men and women.

In Brazil scientists at the National Institute for Amazonian Research are experimenting with "food forests," tree crops that provide food for subsistence farming families. Trees not only shade and protect the soil from erosion, but their roots can go deeper, looking for nutrients. According to David Arkcoll, a scientist at the institute, "the only successful advanced agricultural systems in the tropics are tree crop plantations—oil palm, rubber, bananas. They've worked because they are (1) ecologically stable, viable, sustainable and (2) profitable, so their owners can afford to fertilize the soil. But they don't feed anybody, you see."

Arkcoll points out that fertilizing annual crops does the soil little good because most or all of the added nutrients are pulled out each year with the mature crops. Perennial crops are a better investment, but by a cruel twist of fate most food crops are annuals, and hardly any are trees. However, there are a few tree crops that could be the basis of a stable subsistence farm—plantains, jackfruit, breadfruit (seedlings of which Captain Bligh was carrying aboard the *Bounty* as he sailed from Tahiti on the fateful voyage), and peach palm, to name but a few. For many centuries Indians throughout Amazonia have cultivated peach palms. By selecting individuals with the characteristics they desired, they "bred" more than two hundred distinct types of peach palm. "The Indians made the peach palm," Arkcoll's colleague, Charles Clements, told me. "In Bolivia they developed its oil content; in Yurimaguas they eliminated the spines on its trunk so that they could get heart of palm out of it; in the region on the borders of Colombia, Peru, and Brazil they grew peach palms with high levels of carbohydrates so they could use it as a staple food. Some other varieties have up to 15 percent protein, much more than manioc, which is the staple diet of Amazonia now."

Arkcoll and Clements's experiments indicate that peach palm is easy to grow ("It's practically a weed," says Clements). A small farmer could grow from ten to thirty and get a good yield (about 130 pounds per tree per year) using just household waste. For the first three to five years, before the trees are producing fruit, farmers could grow annual crops in between the trees. Peach palms, according to the variety selected, can be a subsistence crop and a cash crop, producing hearts of palm or oil. It is said to be competitive with African oil palm as an oil producer. In fact the first diesel engine ever made —in France in the 1890s—ran on peach palm oil.

Arkcoll and Clements have devised an interesting way of supplying a peach palm field with the extra nutrients it needs. "We have been experimenting with shallow transportable outhouses over holes of twenty to thirty inches. These can be moved a few yards every few weeks or months and a food tree planted in the old position. In this way a small orchard can be produced around the house in the poorest of soils. The system is comparatively effortless and is also hygienically satisfactory." In spite of its appealing name, the fruit of the peach palm is not particularly delicious. It is not distasteful, merely bland, with the texture, when cooked, of a somewhat mushy potato. Whether the peach palm becomes a staple food in Amazonia or anywhere else will depend on kitchen trials as much as field trials.

In some areas, settled agriculture will always be out of the question. Such land cannot support intensive farming or high population densities. Those who insist that shifting cultivation is an unaffordable luxury because it takes up too much land should answer the question, Compared with what? For many areas in the tropics, shifting cultivation is the only way of farming that works.

8

The End of the Road

Given a choice between highlands and humid lowland rainforests, most people choose to live in the hills. In Colombia, for example, most of the population is crowded into the three Andean mountain ranges that run parallel down the western side of the country. Until recently the great Amazonian forest of the eastern lowlands and the Pacific coast rainforest (the Choco) to the west have been relatively undisturbed. The Pacific forest covers about 5 percent of the country. Although "one of the most poorly known areas on earth," according to Alwyn Gentry of the Missouri Botanic Garden, who has collected plants there, the Choco is well recognized as a great center of species diversity. Botanists believe that it has more endemic species than any other place in the world. With from three hundred to four hundred inches of rain a year and no dry season, it is also one of the wettest places on earth, thanks to the collision of the moisture-laden ocean winds with the Andes.

Isolation is the most effective protection a rainforest can have, and a road, no matter how primitive, is the end of isolation. Already, where roads have penetrated the Choco, there is terrible devastation. The inaccessibility of the Pacific forest, which has protected it for centuries, is now coming to an end. Road building is a boom industry in Colombia. The government wants to encourage people to move out of the crowded Andean region to the eastern and western forests to make a living by logging, crop growing, and cattle raising. There are a few government-sponsored settlement programs, but they are expensive and slow, requiring advance planning, soil mapping, and financial and technical support for the colonists. The favored means of promoting colonization of the frontier is simply to build roads into it and wait for people to use them.

The Cauca River flows from south to north on the Pacific side of the central range for almost the whole length of the country. In the early 1950s a special agency, the Cauca Valley Authority (CVC), was established to protect the river valley and its watershed from the effects of colonization and deforestation. The authority thus has the unenviable task of trying to mitigate the effects of the government's road-building program. With Efren Varela, a forester with the CVC, I followed one road, not yet completed, as far as it had penetrated into the forest.

Setting off from Cali, ninety miles from the Pacific coast, we drove south on the Pan-American Highway through the valley, where the soil and the living appeared equally rich. The highway was lined with flowering hedges; behind them nestled private country clubs. I caught glimpses of long-haired girls diving into the clear blue water of swimming pools and uniformed workmen pushing heavy rollers over grass tennis courts. A new white building, designed to the highest standards of Californian high-tech, user-friendly, corporate architecture, housed labs for Merck and Company, the American drug firm. Farther back from the highway, horses and cattle grazed on lush-looking pastures.

After fifteen miles we turned west off the main road and headed into the foothills of the western range. The paved road became a dirt track switching back and forth across steep slopes. Around us the ridges and valleys were bare and dry, their soft brown color broken only rarely by the green of an isolated clump of trees. On the horizon not a single tree stood up to spoil the smooth line of the bare gray hills. We saw no signs of life, no crops or grazing animals. On some hillsides the sparse ground cover had crumbled away completely under the pressure of rain, leaving the heavy red soil exposed and vulnerable to further erosion. The authority had made some attempts to forestall landslides by embanking the roadsides with rocks held in place with chicken wire. "If remedial forestry measures are not taken immediately, large parts of Colombia (already less productive than in the past), may be rendered uninhabitable in the future because of desertification." This prediction, in a 1980 report by the U.S. Department of Agriculture, seems here to have come true already.

These once-forested lands were cleared for pasture. They belong to wealthy landowners whose main properties are in the valley. The soils here are not suitable for pasture, but the land is cheap, and there is no shortage of peasants willing to clear away the forest in exchange for a season of growing crops before the cattle are brought in. Many of the pastures are abandoned; others are still in use, but with land so poor that each cow needs twelve acres or more to feed on, it is rare to catch a glimpse of an animal.

Varela is worried by the erosion apparent in the deep gullies that run down the length of the cleared foothills. "We want to reforest one hundred thousand acres of eroded soils with fast-growing species like eucalyptus and leucaena that are good for firewood. The aim is to protect the watershed but also to provide some direct use to the people so that they will protect the forest."

Some of the big ranchers are now planting pines to take advantage of generous government credits and tax incentives. Varela, however, has reservations about conifer plantations. "In my opinion," he said, "pines don't do as good a job in protecting the watershed as native forest. You don't have to be a forestry genius to see that the natural forest has greater density, more complexity, more biological activity, and therefore is more biologically stable than a uniform pine forest, which has less biomass, less complexity, and less ecological equilibrium. But these soils are so damaged we can't hope for the native forest to return, so we have to rely on pines, which at least are better than bare land."

After some miles of this empty, degraded landscape, peering down the steep hillside, we see below the road a tiny farm of about two acres precariously laid out on a forty-five-degree slope. It seems a crazy place for a farm; the whole setup looks in danger of sliding down into the valley, but, according to Varela, this farm has been here for at least ten years. The owner, who lives in Cali, is one of the richest farmers in the region. The farm is evidently well tended. "They use good farming practices here," remarks Varela, "intercropping and so on." Banana trees shade rows of coffee bushes, and the areas in between are planted with ground crops, which help the soil to retain its moisture, especially during the dry season. Nature has helped. "Most of the soils we've passed are iron-rich oxisols, what we used to call laterite, impossible for agriculture. But here the soil is quite good; we probably have here a little pocket of volcanic soil."

We drive on, passing several farms, none of which seem to have the order or the luck of the first one. Illegal fires rage where settlers are clearing new land; are these, I wonder, the same men whose abandoned plots we passed earlier? Most of the *campesinos* here cannot live off the land alone; they also hire themselves to land speculators, who pay them to clear more land for pasture or—irony of ironies—to fell native forest so that they can reforest with tax-deductible pines. This is illegal, but the CVC doesn't have the staff to patrol every acre of the thousands of square miles for which they are responsible.

The little hamlet of Villa Colombia is the last town on the road. It has a general store, and we stop there to buy something to drink. On our way out of town, we pass another car, the first we have seen since we turned off the main road. Both drivers stop as a matter of course. It would be as unthinkable to drive straight on past a fellow motorist on one of these little roads as it would for two hikers on an isolated mountain trail to pass one another without a word. The other car turns out to be driven by Alonso García, the CVC manager for this area.

García is responsible for the Farallones National Park, a 360,000-acre forest just a few miles to the west. Except on its outer fringes, it has been isolated and virtually untouched by man. The few naturalists who have been in the area say that it shelters the only bear that is native to South America, the spectacled bear, a species that is disappearing rapidly as its habitat is destroyed. A road connecting the coastal city of Buenaventura with Cali skirts the northern edge of the park. From it and the Pan-American Highway itself, many small feeder roads, including the one we are on, are closing in on the park from the north and east. Around Buenaventura, Carton de Colombia, a subsidiary of the Container Corporation of America, has a concession to log 120,000 acres of the Pacific rainforest; it pulps the trees to make cardboard boxes and writing paper. For part of its route, the road follows the course of the Río Anchicaya, a name well known to connoisseurs of Mother Nature's revenge. It was on the Río Anchicaya in the 1960s that a multimillion-dollar hydroelectric dam silted up within fifteen years of completion, due to severe deforestation of its watershed.

García spoke of the failure of the CVC, an organization of scientists and technicians, to counteract the machinations of local officials and other, more politically oriented agencies. "The CVC is supposed to control the watershed, but we can't control the Department of Public Works. We can't even get them to agree not to build roads on land that is too steep or that is not suitable for colonists. That road" —he pointed to a track behind us—"the Department of Rural Roads built without even consulting us. The first we knew about it was when we saw the bulldozers working.

"Rural people like roads. If the CVC says no, they go to the politicians. They are pushing hard for this road to go right through the park. There are votes in it, but it will be a disaster when the road enters the park, not only for the forest and the animals in it, but for the people living out here. They don't realize that these forests are protecting their land and their water supplies." How was he, I asked, going to protect the park once the road went through? "Well, we have six wardens and we're building two watchtowers. But we have six hundred square miles to cover. We can't keep people out, so our plan is to try instead to minimize the damage. Maybe we can find a way for them to develop a stable agriculture. Maybe we can subsidize a cash crop, or . . ." His voice trailed off. He shrugged and sighed.

One cash crop is already having a drastic effect on the whole region. On the other side of the mountains, coca, the raw material of cocaine, is an up-and-coming, though illegal, crop. The Indians of the Pacific coast have long grown coca for their own use. It is native to these rainforests and thrives on steep, eroded soils on which almost no other plants can survive. The new spate of road building has opened the forest to settlers—and buyers—and given coca growing a big boost. Already coca is the main crop on about twenty-five thousand acres of former forest around the small town of Naya, on the far side of the western range.

Much to the worry of CVC officials, the Department of Public Works is building two roads from Naya across the mountains to the Cauca Valley. García and Varela fear that coca will spread to this side of the mountains, with disastrous consequences for the forest. "People here are hard-working, but coca (you call it snow in the United States, yes?) is a very easy way to make money. Almost every kid in Naya has ten thousand pesos [about $200] in his pocket. Already people are stealing things from here to sell for two or three times as much over there. Two weeks ago six horses were stolen; they'll fetch a price four or five times higher in Naya."

Their fears are not unrealistic. In the 1960s the United States government helped Peru "develop" the rainforest of the upper Huallaga River along the eastern Andes by building roads into the forest. The idea was to move in landless peasants who would convert the forest to agriculture. What the authorities did not foresee was that coca would be the principal crop and that the upper Huallaga would become one of the world's major coca-growing regions, dominated by international drug syndicates and immune to governmental interference.

There are two main landowners along the road we were traveling. One has commandeered a large area of flatland, a plateau amid the mountains. Except for a few scraggly trees and charred stumps, the land is bare. A stream winds through the rough grazing land, and a rough ranch house stands near a small waterhole, but I can see no more than twenty cattle on a pasture of perhaps one hundred acres. The owner is a businessman in Buenaventura, one hundred miles away. For him this ranch is a profitable investment in the short term, a tax deduction, and a status symbol. For the CVC it is the forest destroyed, the watershed threatened.

The second big landowner here controls the surrounding uplands. He, too, is an absentee landlord, employing peasants to clear and work his land, extending his holdings to the west as the road progresses in that direction and abandoning his earlier pastures as they become useless.

As we went on, we saw more forest and fewer signs of human interference. Around one bend, however, we overtook a barefoot man carrying a large scythe. His sunburned, wrinkled face and bony frame made him look old and frail, but his voice was robust. He was on his way to weed his plot, a hilly piece of land above the landslide that bordered the road. He had only recently cut and burned the forest to plant his crops. Did he intend to stay and cultivate the land he had cleared? "No, this land is not so good. When the road goes further, I will follow it to better land." Varela looked at me in despair.

Where the road petered out to nothing at the top of a ridge, about six thousand feet above sea level, we found a solitary house, its mud-caked walls topped by a corrugated iron roof. Samuel Herrera, his wife, and their four children, three girls and a boy, had come here a year ago. From his vantage point Herrera could look down on the property of the businessman from Buenaventura, who is his patron. He was clearing land for this man and at the same time trying to grow crops for his family on an incline too steep for even the apparently goat-footed Colombian cattle to graze upon. Though Herrera was in the middle of his chores when we pulled up, he stopped working to answer our questions. Señora Herrera gave us coffee, weak and very sweet, as the local people take it. The house, neater and more comfortably arranged than I would have imagined possible in the circumstances, had a separate kitchen and an outdoor scullery with running water, ingeniously diverted from a mountain stream by means of a Rube Goldberg system involving rubber tubing, wooden troughs, and an old gasoline barrel. Two of the girls, aged about eight and eleven, helped their mother with the coffee and then played on the bare red clay that was both floor and garden.

Before coming here, Herrera had been working for a large landowner to the south. He left because, although there had been no violence as yet, tension was rising between the settlers and the Paez and Guayambo Indians, whose traditional lands had been invaded. He was growing maize, some coffee, and a few other vegetable crops. The nearest market is in the valley, a seven-hour walk away, he said, but he doesn't go there. Instead, a middleman makes the trek uphill, buys Herrera's produce, and loads it onto a horse for the long journey back to market. Life is clearly hard here; the soil is not favorable to agriculture, the land is so steep that simply moving around the small family plot is tiring; the closest thing to a village is a hard day's walk away; the weather is hot and humid, landslides are common, and the water supply is erratic. What, I asked Herrera, could be done to make his life better? "Build the road further," he said quietly.

9

Partners in Trade

Logging is the main cause of degradation of tropical rainforests. It also earns more money from them than any other activity. In 1980 roughly 4 billion cubic yards of wood were felled throughout the world. More than half of that was burned to provide heat and power. The rest, about 1.8 billion cubic yards, was "industrial wood," used for construction or converted into paper and cardboard boxes. Only 20 percent of the world's industrial wood comes from rainforests, but more than half of that is exported to the richest nations. Thus, virtually all the hardwood logs and more than half the hardwood sawn timber in world trade comes from rainforests. In 1979, for example, 58 percent of the world production of hardwood logs and 75 percent of all log exports came from Malaysia and Indonesia alone. The developed countries, which produce 80 percent of the world's industrial wood, keep almost all of it and import much of the rest of the world's harvest as well. Japan alone takes more than half and Europe more than one quarter of all wood exports.

The rainforest countries of Asia and the Pacific export 70 percent of their industrial wood, half to Japan and most of the rest to countries (mainly Korea, Singapore, and Taiwan) that process the logs and immediately re-export them to North America, Africa, and the Middle East. West Africa exports just over half of its harvest, largely to the EEC; Latin America exports less than 10 percent of its harvest, mostly to North America and the EEC.

Twenty-five years ago world trade in wood was a very different picture. Between 1961 and 1975 developing countries more than doubled the amount of industrial wood they harvested each year. They also exported more and more of what they produced—one third in 1961, half by 1977. In some countries the leap was much more spectacular. Indonesia, for example, exported 148,000 cubic yards of logs in 1961; in 1979 the figure was 167 times higher, 25 million cubic yards. The industrialized nations have been increasing their imports; they now use sixteen times as much tropical hardwood as they did in 1950. Tropical countries themselves use only about twice as much hardwood as they did in 1950.

Most wood must be processed before it can be used, turned from raw logs into lumber, plywood, veneer, or pulp. The industrialized countries process their own wood. Few developing countries have the equipment and skills needed to process more than a tiny fraction of their annual log harvests. Three quarters of their exports, therefore, are logs, less valuable than processed wood. In 1961 half the wood exported from Latin America was in the form of raw logs; now all of it is processed. But Africa and the Asia-Pacific region are still exporting the same low proportion of processed wood—35 percent and 20 percent, respectively—as they did in 1961.

Wood exports bring around $8.7 billion a year to the developing countries. Asia and the Pacific get 70 percent of that; most of the rest goes to Africa. Indonesia gets more than 20 percent of its gross foreign exchange earnings from wood, as do Gabon, the Congo, and Burma. In Malaysia, Cameroon, and the Ivory Coast, wood exports provide from 10 to 20 percent of gross foreign exchange earnings. But net foreign earnings are often much lower. Deducting the cost of imported equipment, the income that foreign workers send home, and the profits that foreign companies transfer to sister companies elsewhere often cuts wood export earnings in half. Indonesia's net earnings from timber exports in the early 1970s, for example, were only a quarter of its gross earnings, though by the end of that decade, when loopholes had been tightened, net earnings had doubled to 50 percent of gross earnings.

Since 1975 Jant, a subsidiary of Japan's Honshu Paper company, has been clear-cutting a 330-square-mile area of rainforest in Papua New Guinea. The entire area will have been deforested before 1990. According to the Jant prospectus, "Every day one hundred various types of heavy vehicles and their operators are working and making their presence known in the green jungles of the Gogol area." The company was supposed to reforest as it cleared the area, but it is logging ten times faster than it is replanting. When the first concession is used up, Jant can exercise its option on another 260 square miles nearby. Every month it ships twenty thousand tons of wood to the Honshu paper factory in Japan. Jant pays no dividends or income taxes in Papua New Guinea because it sells the wood to its parent company at such low prices that it never makes a profit. That practice, called transfer pricing, is said to cost Papua New Guinea $11 million each year.

In any case, for all but a few countries, the income from exporting mostly unprocessed wood is eaten up by the need to import sawn timber, plywood, and paper. Many tropical rainforest countries import more than they export, mostly in the form of paper and paperboard. Nigeria was once a major timber exporter. It is now on the verge of complete deforestation and imports a hundred times as much wood as it exports. Exporting unprocessed wood cheaply and paying high prices for processed imports is so wasteful that some countries—notably Malaysia, Indonesia, and the Philippines—have tried to ban the export of logs. None, however, has succeeded in imposing a complete ban, partly because Japan and Europe, the major importers of tropical wood, prefer to buy logs cheaply and process them themselves. They tax processed wood in order to discourage its import. The EEC, for example, which buys 70 percent of Africa's wood exports and 40 percent of Asia's, allows free trade in unprocessed logs, but puts an 11 percent tax on processed timber.

Timber operations and other rainforest developments, such as oil and gas drilling, mining, and ranching, should contribute to national economies by paying taxes and royalties and by providing jobs. But they may contribute very little, or even cost the country money, while making a profit for the owners. One study of U.S. subsidiaries in Latin America found that over one three-year period they financed 78 percent of their investments with money borrowed locally, thus reducing the amount of money available to local investors. The same companies sent 52 percent of their profits home, rather than investing them in new projects in Brazil. In 1974 a survey by the Brazilian government of 115 multinational companies revealed that they imported $2.1 billion worth of products and equipment more than they exported. A United Nations study showed that the $434 million invested by multinational companies in Indonesia's logging and processing industry between 1968 and 1974 created only seven thousand extra jobs for Indonesians. One of the most common criticisms of foreign companies is that they do not train local people for senior positions.

Brazil encourages companies to invest in the Amazon by offering them a 50 percent income tax rebate on their investments elsewhere in Brazil, tax holidays of up to ten years, loans with negative interest rates (in real terms), and exemptions from sales tax and import duties. It has proved to be an expensive way of making jobs. Each industrial job created costs the government $34,000; each ranching job $63,000. As for bringing in foreign capital, half the cost of the industrial projects and 72 percent of the cost of the livestock projects were covered by fiscal incentives and other subsidies—that is, paid for with money diverted from the national treasury. Much of the rest was borrowed from Brazilian banks. Since 1979 new ranches in the Brazilian Amazon have no longer received such support because, in the words of the environment minister, Paulo Nogueira Neto, "Experience has shown us that ranching projects in the rainforest are a disaster." But the incentives continue for existing ranches.

In business negotiations with developing countries, the two hundred or so giant multinational corporations enjoy a decided advantage in terms of financial resources, technology, managerial expertise, and influence. In 1981 Exxon, the biggest of these multinationals, had an annual revenue bigger than the gross national products of at least sixty-four countries. Unilever was the twenty-fourth largest company in the world in 1981; it had a total revenue of $24.1 billion. The Solomon Islands had a gross national product that year of $95 million. Unilever employs 320,000 people directly and another 50,000 through associated companies. The population of the Solomon Islands is 220,000. Unilever's logging operations there employ 450 people and account for 15 percent of the value of all the country's exports. Unilever's contract with the government of the Solomon Islands does not require it to reforest logged-over areas. The contract between Jant and the government of Papua New Guinea obliges the government to pay half the cost of reforestation, which is estimated to be $180 per acre, although Jant pays royalties of only about $40 per acre.

Although reforestation is a normal part of the logging cycle in North America and Europe, many of the companies that operate in rainforest countries feel that replanting is a government job. Concession holders say their leases, usually about fifteen or twenty years, are too short to offer an incentive to replant. They prefer, as a rule, to take their quotas as quickly as possible rather than to establish long-term operations in a climate of political and economic uncertainty. Governments, according to Unilever's forestry director, Don McNeil, "have—or should have—the right people to do the job. They should be doing reforestation on the ground close behind the commercial extractor. It is in everybody's interest to grow more trees."

Many countries are now trying to attach stricter conditions to the logging concessions they grant, so as to encourage reforestation or domestic processing. Since 1979 Indonesia has required its concession holders (of which there are at least five hundred) to deposit $4 for every cubic yard of wood they remove, to be refunded as they reforest the logged-over area. However, only a handful have begun a reforestation program. The rest are forfeiting the deposit and passing the problem back to the government. Unfortunately, reforesting logged-over rainforest is still an experimental matter, and neither the government nor private contractors have the resources or experience to plant even a small part of the huge area cleared every year.

In Papua New Guinea local enterprises have only one fifth of the forest concession area; the rest is leased to foreign companies. According to the United Nations Center for Transnational Corporations, "transnational corporations or their affiliates account for more than half of the international commercialization of the hardwood resources of Asia and the Pacific." But in many countries local investors dominate forestry and ranching in rainforests, though the relatively few large multinational companies are more visible. Even local investors, however, can be outsiders in terms of the forest, argues Brazilian ecologist José Lutzenberger. "What's happening in the Amazon is just another form of imperialism whether the people going there have a Brazilian passport or a foreign passport. They are exploiting the local resources at the expense of the local population. Brazil is so large that we can afford to leave the Amazonians in peace, at least for the next several decades. We should leave the Amazon alone and concentrate on the problems we have to solve in the south of the country."

For most items, tropical rainforests provide the industrialized world, not with essential supplies, but with "extras" that are only worth having if the price is right. The trade in most rainforest products is more important to the sellers than to the buyers. This is because, as a rough rule of thumb, the richest 20 percent of the world's people consume 80 percent of the world's goods. Each additional unit is worth less and less to us, since it represents a smaller and smaller proportion of the whole, but as the producing country's raw materials are depleted, extracting each extra unit becomes more and more disruptive.

In the industrialized world as a whole, the average person uses 275 pounds of paper a year. In the United States alone, where a single copy of the Sunday *New York Times* weighs four pounds, each person uses 650 pounds. The average person in a developing country uses only 15 pounds of paper a year. Although developed countries buy more than half the industrial wood produced from tropical rainforests, those imports account for only about one tenth of all the wood and paper used by the industrialized world—and less than 2 percent of the huge American demand for wood and paper.

Though it imports more wood than all other countries combined, Japan itself has 60 percent of its land under forest, more than any other industrialized nation outside Scandinavia. Its forests contain enough wood to replace its imports from Southeast Asia and the Pacific for 150 years, at current rates. But the Japanese harvest their own timber at a rate far below what their forests could support. In fact their annual harvest today is only two thirds of what it was in 1966. During the same period imports of tropical logs more than doubled. "Japan has a very clear strategy," according to West German forestry expert Professor Hans Steinlin. It is "to protect its forests for as long as possible, although it means overexploitation of Southeast Asia and the Pacific region." Japan is now becoming so worried by the imminent depletion of the Asian and Pacific rainforests that its pulp manufacturers have established a fund to finance plantations of fast-growing timber species, such as Caribbean pine, in selected countries. Plantations, should they succeed, may provide a reliable supply of timber; in terms of wildlife, useful plants, and ecological protection, they will not replace the lost rainforests.

. . .

In the Indian state of Bihar, tribal people have been burning down the teak plantations established by the government in the place of the natural forest, from which local people have long gathered medicines, fruits, timber, and firewood. Plantations, no matter how valuable the wood grown in them, do not provide this variety, nor are they open to the people. When the trees are mature, commercial loggers pay the state for the right to fell them. This is a classic instance of wealth being redistributed upward: A permanent asset, the benefits of which are broadly distributed among many people, is converted to a short-lived one that profits the few with the money to exploit it. The plantations are an attempt to counter the devastating ecological effects of deforestation in the Himalayas. But just as hard on local people is the social and economic disruption that deforestation causes, and in that respect timber plantations cannot take the place of communally held natural forests.

The roots of the problem are in well-meaning attempts by British colonial foresters in the late nineteenth century to protect forests from the local people who had for centuries collected firewood, cut wood for buildings and plows, and grazed their animals in them. The people had their own conservation practices, encouraging the regeneration of useful trees and voluntarily limiting their grazing and collecting. But the foresters decided to turn some of these multi-use forests into productive timber stands, and declared many forests off limits to the people. Apart from the hardship caused by the loss of a large part of their income, local people deeply resented what they regarded as the theft of their communal property. This resentment culminated in deliberate fire setting during 1916 and 1920-21, when the people destroyed tens of thousands of acres of "stolen" forest.

The forestry service could not enforce the forest closure laws absolutely. In many areas people were able to enter the forests as before—with one difference. Now they knew the forests did not belong to them; the trees would eventually be sold to the highest bidder, in all likelihood a stranger to the region. People abandoned their time-honored conservation practices in an effort to get as much of the forest as they could before it was barred to them or felled by outsiders. In areas where the closure was effective, the local community had to satisfy all its needs from the small village forests the authorities had not claimed. The result is that both the village forests and many of the state forests have been overexploited and degraded, in many cases to the point of complete ruin.

Now, having experienced twenty-five years of disastrous flooding and a dire shortage of fuel wood, villagers are beginning to band together to protect the remaining forests and plant new ones—forests for local use, not commercial plantations—where the slopes of the Himalayas are bare. The current campaign originated in 1973 near Gopeshwar, close to India's border with Tibet, when the Forestry Department banned a village carpentry cooperative from using wood from the local forest and instead issued a felling license to a firm from Allahbad, hundreds of miles away. The villagers, who had recently suffered several years of terrible flooding that they attributed to government-licensed deforestation, decided to protest. In March, when the company sent its loggers in, a local leader named Chandi Prasad Bhatt declared: "Let them know they will not fell a single tree without felling one of us first. When the men raise their axes, we will embrace the trees to protect them." The trees were spared and the Chipko Andolan (the Hugging Movement) was born.

Chipko has spread throughout the region, largely supported by women, who know firsthand the hardship of having to walk for miles to find a few sticks of kindling. In one village, according to Indian journalist Anil Agarwal, the women did not speak to the men for six months, until they agreed to fight the government's plan to convert their forest to farmland. The men had been promised jobs and were willing to accept the change, but the women objected. Ultimately they persuaded both their men and the government to drop the idea. Chipko is campaigning for a ten-year ban on felling, to which the state governments, which receive royalties for every felled tree, are bitterly opposed. Chipko workers have also planted hundreds of thousands of saplings on the bare slopes of the Himalayas.

The people want forests they can use. The official planting programs emphasize eucalyptus plantations. In August 1983, several thousand villagers marched on a government tree nursery and uprooted the eucalyptus seedlings, replacing them with tamarind seeds. They offered to cooperate with the Forestry Department in planting other species. The police were overwhelmed when 1,500 people gave themselves up, so no one was arrested. One farmer explained, "Thirty years ago we had thick mixed forest, then some ten years ago eucalyptus was introduced to our area. For the first three to four years the eucalyptus plantation was fenced and guarded and we were not allowed to graze our cattle, Then after four years the eucalyptus had grown to such an extent that there was no fodder under the eucalyptus tree. Nor would the cattle eat the leaves." Another problem is that eucalyptus trees grow with a straight stem without side branches that could be used as fuel. Instead the villagers are forced to use as fuel the cow dung that would have fertilized their crops. Thus, eucalyptus plantations deprive local people of fodder, fuel wood, and fertilizer, not to mention the wild food and medicine to be found in a mixed forest.

Sunderlal Bahuguna is an old man who has walked many miles back and forth over the Himalayas to spread the aims of the Chipko movement. In 1982 I heard him speak about his work: "The first product of the tree is soil, water, and oxygen. They say the trees are unproductive until they are cut; that the land is needed for development. What is development? For most of us development means affluence. You have to produce more and more things for your luxury and comfort. Ultimately this tells on your resources because everything has to come from nature. You have to decide whether development means affluence or whether development means peace, prosperity, and happiness. I think peace and happiness have gone away for the sake of prosperity."

10

Trees for the Wood

Five hundred years ago, while the Philippines was still unknown to Europeans, rainforests covered almost all its eleven thousand islands. By 1900, plantations of sugarcane, tobacco, and hemp established by Spanish traders had destroyed about one fifth of the islands' forests. In 1953, after fifty years of American occupation and a great expansion of the plantation economy, just over half the country was forested. Today, although the Philippines Bureau of Forestry Development claims that "forestlands cover about 54 percent of the total land area of the country," satellite pictures commissioned by the Philippines Department of Natural Resources show that in fact less than one third of the country is forested. The Philippines now has under three quarters of an acre of forest per person, the lowest ratio in Southeast Asia.

Every accessible forest in the Philippines that contains useful timber has been leased to timber companies. For many years after the Second World War, the Philippines exported more logs than any other tropical rainforest country. But by the early 1970s its forests were so depleted that it was impossible to maintain exports at such a high level. Between 1977 and 1980 the number of timber licenses declined by one third to 261, a drop that is indicative of depleted forests rather than of a strict licensing policy. President Marcos, worried about the ecological and financial implications of national deforestation, imposed a ban on log exports.

The Philippines is, or was, especially rich in the most commercially valuable type of tropical rainforest, dipterocarp forest. Dipterocarps are a family of exceptionally tall trees that grow only in Southeast Asia, often in relatively homogenous stands. Like teak forests elsewhere in Southeast Asia and mahogany forests in Latin America, dipterocarp forests are an exception to the general rule that tropical rainforests are a very diverse mixture of species. They are among the finest tropical timbers—well grained, hard, and durable. But because there are almost six hundred species in the family and trade names vary from country to country, neither their common nor their scientific names are well known outside the trade. Nonetheless, most North Americans and Europeans have at some time or other walked on, sat at, or leaned against Philippine mahogany, meranti, lauan, mersawa, or another dipterocarp. The consequence of their popularity is that all the accessible dipterocarp forests in the Philippines, and in most of the rest of Southeast Asia, will have been logged out by the middle of this decade. The government's own Five-Year Plan 1978–1982, acknowledges this: "By the early 1990s when the virgin forests in Mindanao are completely logged off, production will reach the limit of the allowable cut of about 13 million cubic meters. This will decline further to about 12 million cu meters by mid-1990 when all Luzon virgin forests are also completely logged off."

Most of the Philippines' remaining forests, dipterocarp and otherwise, are on the island of Mindanao, the southernmost and second biggest island in the archipelago. Mindanao also has PICOP, Paper Industry Corporation of the Philippines, the biggest producer of wood and pulp in Asia and by reputation the most enlightened tropical logging company in the world.

. . .

I had been in PICOP territory for some time before I realized it. Only when the bus came to a checkpoint and we were ordered out by guards holding rifles did one of the other passengers explain to me that we were already in the concession. I was not eager to be searched, since my bag was stuffed with what the government had branded subversive literature—information about deforestation and treatment of tribal minorities—given to me by priests and Catholic layworkers elsewhere in Mindanao. I need not have worried; I was allowed to stay on the bus—whether because I was a woman or because I was a foreigner I do not know—while all the men were checked. The two things the guards were looking for, a PICOP security official told me later, were weapons and mimeograph machines. All mimeograph machines must be registered with the Bureau of Communications. A signboard proclaimed the name of the place as Cassandra Crossing. When we were finally waved through, the first thing we passed was a funeral procession. A little boy carrying a cross led the way. Behind him came two men holding up a tiny wooden coffin.

There was one more checkpoint, again with armed guards searching the bus and passengers, before we reached Bislig, the town on the Pacific coast where PICOP has its offices and factories. I waited outside the fenced-off compound while yet another armed guard checked my credentials with the head office. Finally a PICOP chauffeur appeared and whisked me off to a meeting with the public relations director, Carlos Andreas. After all that security, I was left alone in Andreas's office, waiting for him to return from another appointment. I occupied myself innocently counting the framed awards for outstanding civic responsibility, acknowledgments of corporate gift giving, and certificates of good safety records that smothered the walls. When I had counted twenty-seven on one wall and got up to sixteen on a second, Andreas came in. I asked him about PICOP.

. . .

The company was founded in 1952, a joint venture between the American firm International Paper Company, the world's largest paper producer, and the Andrés Soriano Corporation. Soriano is the only Philippines-based multinational company and one of the very few headquartered in a developing country. It is, for all the nationalistic rhetoric, in one sense an American company; the members of the Soriano family who still control it are U.S. citizens. Its fortunes are founded on San Miguel beer, though its interests now include mining, real estate, and retailing. Soriano's founder, Don Andrés Soriano, died in 1964, but even in his lifetime his knack for success and his much publicized social conscience had made him a Filipino legend. He was an early adherent of vertical integration in industry. Not content with brewing San Miguel, Soriano gradually took over its bottling, distribution, and marketing as well. Wanting also to control the labeling and packaging, Soriano formed the Paper Industry Corporation of the Philippines.

PICOP has permission to log more than seven hundred square miles of exceptionally rich virgin forest. Ninety percent of the trees are dipterocarp species collectively known as Philippine mahogany. The entire concession (apart from a few inaccessible areas that are being left alone) will have been cleared by 1987. One third is being clear-cut and converted to plantations of fast-growing trees. The company is logging the rest in a way that it hopes will permit regeneration and a second cut in thirty-five years.

There are two main ways of taking trees out of a forest, clear-cutting and selective logging. Under the latter system only a proportion of the total number of trees is removed, the theory being that the forest is so little disturbed that it will regenerate naturally and in time be ready for another cut. "Selective logging is good in theory," Angel Alcala, professor of biology at Silliman University in the Philippines, told me, "but it does not really work. With selective logging, you are supposed to take only a few trees and leave the rest to grow, so that you can return later and take some more, without destroying the forest. This is supposed to be a sustainable system. But here, although they use the phrase selective logging, there is only one harvest, a big one. After that no more."

The reason for this is that if a company takes out only 10 percent of the trees in an area, most of those remaining are often severely damaged by the way the loggers operate. The most disruptive thing they do is to construct roads and tracks. One study found that 14 percent of a logging area is cleared for roads, and another 27 percent for skidder tracks (on which logs are dragged out to feeder roads). No tree to be felled can be farther than eighty feet away from a skidder track, since that is as far as the chain can reach. In the rainy season, a skidder track can be used only two or three times before the chains or tires destroy it, so new tracks are constantly being opened. Thus more than 40 percent of a concession can be stripped of protective vegetation and highly liable to erosion. In addition, heavy equipment, such as bulldozers and tractors dragging immense trees, goes back and forth over these tracks. Their weight compresses the soil, forcing out the air and making it difficult for vegetation to recolonize the area. This damage can be reduced by using chain saws to do some initial cutting up while the logs are still in the forest, and by pulling them to the main roads with buffaloes or elephants instead of machines. A Filipino company that tried this system found it economically sound. It also provides more jobs, since it uses four to five times as much labor as the more damaging mechanized method.

Researchers have also found that in dipterocarp forests with an average of fifty-eight trees per acre, for every ten that are deliberately felled, thirteen more are broken or damaged. Other studies have reached the same conclusion: Selective loggers damage more trees than they harvest. In one Malaysian dipterocarp forest, although only 10 percent of the trees were harvested, 55 percent were destroyed or severely damaged. Only 33 percent were unharmed. The logging manager of the Georgia-Pacific concession in Indonesia told me that they damage or destroy more than three times as many trees as they deliberately harvest. The others are butted by bulldozers, lose their branches to falling trees, or are brought down by the pull of tangled vines. The problem is made worse by the ease with which in the tropics a broken limb or ripped bark can become infected. Research by one timber company in the Philippines found that at least half the trees left standing after harvesting are rotten inside as a result of logging damage, although from the outside they appear to be healthy.

In North America and Europe, where logging is now more strictly regulated and timber companies know they must rely on the same concession for future supplies of wood, trees are harvested so as to minimize residual damage. In the northwestern United States logging companies sometimes use helicopters and balloons to lift trees out of forests. Another technique is directional felling, marking and cutting a tree to fall where it will do the least harm. A study in Malaysia showed that cutting the lianas that bind trees together could reduce logging damage by 20 percent, while adding less than 1 percent to the price of exported logs.

Critics also claim that loggers tend to take too many of the most vigorous and well-formed trees. Degrading the breeding stock will thus eventually result in a degenerate forest. And regardless of the quality of the trees left behind, a forest that is too heavily logged may never recover. This is particularly a problem in the dipterocarp forests and others that are dominated by commercially desirable species. In the highly diverse rainforests of tropical America, loggers take an average of approximately 4 cubic yards of wood per acre. In African rainforests the figure is 7.1 cubic yards. But loggers take from 21 to 52 cubic yards from an acre of dipterocarp forest.

A dense stand of a commercially desirable species may tempt loggers to extract too many trees. But the more common condition —many species, sparsely distributed—also causes difficulties. The market accepts so few species that loggers are forced to cast their nets more widely. From the West African rainforest, where there are thousands of species of trees, only about twenty are exported to Europe. None of the others are acceptable to the market—either because they have technical limitations, or because wholesalers are unfamiliar with them. Indonesia, which is the world's biggest exporter of tropical hardwoods, has only twelve commercial species. When I suggested to Herman Haeruman, assistant minister in Indonesia's Environment Ministry, that the market for tropical timber should be developed to accept more species, he laughed ruefully and said, "Yes, we try to, but the market is outside." In Amazonia only about twenty-five species are used commercially, and fewer than ten are accepted by foreign markets. Loggers only want species that they can sell, so when desirable species are widely scattered, they tend to go farther afield, disturbing a large area of forest for small returns.

. . .

In Mindanao I went into the PICOP concession with Joey Kanapi, a forester and a second-generation company man. His father was the firm's public relations officer until 1981. His roommate at forestry college, Delfin Ganapin, is now the head of the Philippines Environment Federation and a leading campaigner against deforestation. Like most foresters who work in the field, Kanapi loves the rainforest. The bone of contention between many foresters and ecologists is not the value of the forest but the former's faith in technical progress and the latter's caution. The chief logger of a major timber company in Southeast Asia assured me that it is possible to harvest a tropical rainforest over and over again, safely and profitably. "No one's done it yet," he said, "but it can be done."

Driving through the concession, we passed a number of very bedraggled patches of forest, almost bare except for some white-trunked eucalyptus that the company planted years later to fill them out. These are the forests that PICOP logged twenty or thirty years ago. They should now be yielding a second cut, but they were so severely damaged that they have not regenerated as hoped. "The Bureau of Forestry Development [BFD] got stricter about enforcing the selective logging standards in the early 1960s," commented Kanapi as we gazed at a sparse copse. "Before that, selective logging was practically clear-cutting."

Today, however, PICOP has the best reputation of any logging firm in the country, perhaps of any in the region. "PICOP is more sensitive to environmental and human considerations than most in Southeast Asia," Colin Rees, the environmental officer of the Asian Development Bank, told me. The government's Bureau of Forestry Development judges PICOP's performance as "satisfactory, better than satisfactory." And even Alcala had conceded that "PICOP has a good reputation among loggers." Its technical achievements are also widely admired. "PICOP is now moving very close to sustained-yield management. It is one of only a very small handful of tropical forest industrial companies which can even make this claim," Hugh Fraser, the editor of the trade journal *World Wood,* commented after a visit there.

Nonetheless, there are problems. The Bureau of Forestry Development says that the maximum acceptable damage to a logged-over forest, including roads, is 40 percent. According to Eulogio Tagudar, who was for many years the company's chief forester, PICOP's method of logging damages only 30 percent of the trees, but that does not take road and track damage into account. In the field I learned that in truth as many as 60 percent of trees are damaged. A logger explained how they fudge the damage figures. "If there are eighty trees an acre, there are maybe six that we want. But we mark, say, ten trees to be cut. We don't cut the extra four. Instead we take out some of the trees that we have damaged and count them as the extra four. That way it looks like we haven't damaged any, or that our damage rate is very low."

"The concept of selective logging is very good," Kanapi told me. "The trouble is that if you worry too much about protecting the residuals, you get a much lower production."

The government only allows as much wood to be taken from a dipterocarp forest as it believes the forest can regenerate within thirty-five years. The allowable cut formula is based on expectations that the trees left behind will grow four tenths of an inch a year, which they do in areas with exceptionally heavy rain and twenty inches of topsoil. In most areas, however, growth is far slower, sometimes less than half the predicted rate. Where this is true, the virgin forest is being cut at least twice as fast as it should be.

Thanks to a very wet climate, PICOP's forests grow well all year round. And the company takes only about 10 percent of the total volume of wood, roughly seventy-three cubic yards per acre of virgin forest. But the company's measurements of annual growth show that the second rotation will only yield thirty-seven cubic yards of useful wood per acre, half as much as the first cut and not enough to keep the company's plywood, veneer, and sawmills functioning at a profitable level and to keep up log exports. Foresters expect a smaller second cut, but PICOP's is smaller "by a much greater amount than anticipated." Inaccurate estimates of how fast the forest would recover are one reason for the shortfall, but there are others. One is that the logged-over forest started out with smaller volumes than predicted because so many of the residual trees were damaged by the original logging operation. Another is that loggers often overcut on the first round. As a logger in the field told me, "We tend to manipulate the records to show that volume taken that would just pay for the area, but we might have taken more than that."

I asked Kanapi if he was worried about getting diminishing returns on each successive rotation. It was, he said, a problem, but one that could be beaten by changing the extraction rules. At present the government allows PICOP to take out all trees larger than thirty-two inches in diameter, and a certain proportion of those that are twenty-four inches or more in diameter. But, Kanapi argued, if on the second rotation they could take out all trees bigger than twelve or sixteen inches around, "we would have enough." Taking smaller trees on the second cut would not, of course, make the forest grow faster for a third, fourth, and fifth rotation. It would, however, mean a less valuable harvest, since the trees, too small for timber, would have to be used for pulpwood.

As it is, PICOP has been forced to beef up natural regeneration by planting extra trees on all its selectively logged-over areas, or at least those that have not been so overexploited that they are being replanted completely. The main trees are two fast-growing species, *Albizia falcata* and *Eucalyptus deglupta* (known locally as *bagras*). This system of enrichment planting is expected to increase the amount of wood that can be taken out after thirty-five years to 325 cubic yards. Company officials selected the two species, which are widely used as pulpwood, for their fast growth, rather than for any similarity with their noble neighbors, the dipterocarps. But to everyone's surprise, both the fast-growing species can, if left to grow for twenty to thirty years, be used for more valuable products, such as lumber, veneer, and blockboard. *Bagras,* whose grain is very similar to that of the dipterocarps, is especially valuable when older and larger.

Any forest can give a sustained yield—that is, it can be logged over and over again—on two conditions: The logging must be mild enough to allow the forest to recover, and the forest must have enough time to recover between loggings. It may once have been possible for a commercial logger to take low volumes of wood, allow a long period of regeneration, and still make a profit, but today only very small-scale, unmechanized outfits using cheap labor can afford to treat the forest that way. Big timber companies need higher volumes to cover the cost of their more expensive harvesting methods. And they need marketable species. This is a problem for loggers, since, having extracted the species they want from a virgin forest, they cannot be certain—even if they took care to leave behind a large number of "seed trees" of desirable species—how many of the trees they can sell will be there when the forest does regenerate. So the vast majority of logging companies simply take what they can get on the first cut and move on.

This tendency is illustrated in a U.S. State Department review of multinational forestry corporations in the tropics. Brazil once had extensive stands of araucaria pines, but, according to the report, "there have been serious problems in regenerating this species, so the lumber and panel industries will undoubtedly move to other regions and substitute suitable softer broad-leaved species." There is no reason to believe that those "suitable softer broad-leaved species" will recover from the attentions of the loggers any better than the araucaria have done. In Papua New Guinea, a joint venture between the government and the British Inchcape Corporation has heavily logged a vast araucaria forest. There, too, despite brave attempts by government foresters, regeneration is failing. Under Philippine law logging companies must provide for the regeneration of forests they have worked. But a survey in 1975 found that only 17 percent of the licensed loggers in the Philippines did anything toward that end.

. . .

In a very few places logged-over rainforest is regenerating well, due to a combination of careful logging and a favorable climate. Carton de Colombia, which is 65 percent owned by Mobil's Container Corporation of America, has a 125,000-acre logging concession near Buenaventura on Colombia's Pacific coast. The company has been clear-cutting this forest since the mid-1960s, when its engineers developed a way to make pulp from a mixture of as many as 120 different hardwood species.

As a rule, ecologists deplore clear-cutting. It exposes the soil to erosion, leaves behind no source of seeds, and creates a hostile habitat for the forest wildlife. On the other hand, by making use of all the trees in a given area, clear-cutting could reduce pressure to exploit the rest of the forest, though it is only fair to say that this has not been a noticeable effect of clear-cutting operations in general.

Carton de Colombia's method of clear-cutting is perhaps the most advanced in the tropics. Company loggers use chain saws instead of bulldozers to cut trees down, and they lift them out with a system of ropes and winches instead of dragging them out with tractors. They clear only small areas at a time, patchwork-style, within the natural forest, so that the cleared area can be reseeded from neighboring forest and the watershed is not stripped all at once of its protective cover. In the older parts of the concession the secondary growth is impressive. "It looks," says Al Gentry, "like you're getting the primary forest species coming right back."

Clearing the forest at a rate of around 3,000 acres a year, the company expects to have harvested the last of the accessible timber on its concession when the lease runs out in 2004. The original plan was to then go back for a second cut, something which should be possible, given the good rate of regeneration. "But," explained Carlos Jaramillo, a company forester, "we have a very serious problem here, with the colonists cutting the secondary growth before it is mature." Settlers who have followed the logging roads into the forest poach the young trees as soon as they are pole size, big enough to use in house building. It now looks unlikely that Carton de Colombia will be able to get that second cut. Although it still talks about the success of its natural regeneration program, the company has already applied for a new 500,000-acre concession, just north of its current one. In 1980 Weyerhaeuser, the American timber company, pulled out of its investment in one of the biggest Indonesian logging operations. One factor was the fact that, according to company officials, local people were poaching thousands of dollars' worth of wood a day.

In Australia, the state government of Queensland, having in the past allowed most of its tropical rainforest to be degraded by overexploitation, now claims to be logging the remainder in such a way as not to disrupt the forest's ecological balance or its protective functions. But most commercial loggers would not accept the stringent conditions the Queensland Department of Forestry imposes. By strictly limiting the number and size of trees that can be harvested, Queensland loggers get an average of twenty-one cubic yards per acre from virgin forest. On the second cycle, forty or fifty years later, they get twelve cubic yards. Queensland's extraction rates are low compared with the vast majority of rainforest logging operations. PICOP, by comparison, extracts seventy-three cubic yards per acre from its virgin forests.

Finding that erosion after logging increased the amount of sediment in nearby streams by ten times, the Queensland forestry department imposed a complete ban on logging during the three-month rainy season. Most of the sedimentation came from poorly located roads and tracks, the making of which is now more strictly controlled. Queensland foresters mark every tree that is to be cut and indicate the direction in which each should be felled.

North Queensland has set aside 83 percent of its remaining two million acres of rainforests for watershed protection, aboriginal reserves, recreation, and research. So logging directly affects only a small part of the rainforest. The state demands only that its logging operation break even; this it does with its annual harvest of 105,000 cubic yards, in spite of selling logs to industry at artificially low prices. Logging in Queensland is unusually efficient because of the high proportion of commercially usable species, 150 in all. "Before the Second World War," commented Queensland forester Tim Anderson, "there were only a few commercial species in Queensland. We've got to this point by virtually forcing the millers to accept more species."

The harvesting cycle and yield that are appropriate for the Queensland rainforests will not apply directly to other rainforests, but the principle that selective logging should not exceed the rate and degree of a forest's natural flux does. "There is a limit to the buffering ability of the forest," Anderson said, "and selective logging has to lie within that limit. Heavily logged plots have the guts ripped out of them, and that affects the environment of the forest, the nutrient levels, the microclimate, the soil-water balance, so as to inhibit their recovery. Selective logging, by definition, can't go that far."

. . .

Since early colonial days, foresters have been searching for ways to help tropical rainforests regenerate a higher proportion of commercially valuable species more quickly after logging. The foresters' aim was, by making the forests a commercially viable renewable resource, to encourage professional loggers to protect them. In that way, although the forest's natural diversity would be somewhat reduced, its protective functions vis-à-vis the watershed, climate, and wildlife would continue. Foresters did develop a number of systems under which it is possible to log tropical rainforests without destroying them. But none of those systems produce as much wood, or produce it as predictably, as timber companies demand.

In fact less than 5 percent of the world's tropical broad-leaved forests are under any kind of management at all, and 80 percent of those are in India—a remnant of the colonial government's Indian Forest Service. Alfred Leslie, a forester working for the United Nations Food and Agriculture Organization (FAO), carried out a comprehensive survey of rainforest management systems. He notes that "after several decades of experimentation on a fairly large scale, many foresters have concluded that natural forest management systems may be ecologically impossible. . . . Even using social benefit/cost analysis, it is difficult to endorse natural forest management as an economic proposition." More and more foresters believe that only plantations of fast-growing species will meet the demand for wood from tropical rainforests.

Is it technically possible to establish such plantations? Many foresters are optimistic. John Spears, the forestry advisor to the World Bank, argues that there has been enough success with such species as eucalyptus, gmelina, and pine "for us to be fairly certain that plantations of these species can be successfully grown in much of what was formerly wet tropical forestland, provided care is taken over soil and site selection and crop management. The arguments in the 1970s against such intensive plantation development have been mainly voiced by ecologists whose concerns, quite rightly, have been that wholesale conversion of large areas of wet tropical forests to plantations would imply a permanent loss of a unique eco-system."

If the natural forest cannot regenerate, plantations are the best way to rehabilitate degraded soils. The canopy protects the soil, and the deep tree roots retrieve nutrients that shallowly rooted plants cannot reach. Plantations also have a much greater yield than rainforests. On average, undisturbed rainforests grow at the rate of about 1 cubic yard per acre per year. Logged-over forests grow somewhat more quickly, due to the extra light allowed in when part of the canopy is removed; they increase at an average rate of 3.3 cubic yards per acre each year. But an acre of falcata grows almost six times as fast, about 18 cubic yards in a year. Unfortunately, there are now fewer than twenty million acres of industrial (as opposed to agricultural) plantations, and they are being established at only one tenth the rate that natural forests are being destroyed, and outside Latin America almost all plantations have been established by governments rather than by private companies.

Another unfortunate fact is that investors would rather cut down virgin rainforest for plantations than establish them on abandoned, deforested lands. There are two reasons for this: The first is that cutting and selling the valuable rainforest species helps to defray the cost of establishing the plantation and brings money in during the years before the plantation is ready to be harvested; the second is that open land is likely to be occupied. By clearing dense forest, plantation owners face fewer conflicts over land rights.

. . .

In 1971 PICOP started plantations of fast-growing trees, mostly *bagras,* on land that had been clear-cut. "We foresters are against clear-cutting because of its effect on the ecology," Kanapi told me, "but with the demand of these pulp mills, the natural forest just can't supply it." The company plants almost 8,000 acres a year, and is about two thirds of the way to its target of 150,000 acres of plantation. It began harvesting in 1980 and now cuts from 1,000 to 1,250 acres a year.

The biggest problem in establishing a plantation, according to Tagudar, is weeding—especially where, as at PICOP, there is no dry season, when weeds might die back naturally. It costs PICOP about $400 per acre to plant the seedlings and get them past the first vulnerable years. There are about five thousand acres of "fail spots," where plantations have died for reasons no one knows. "We are looking for species that will grow on those soils," Kanapi said, "not necessarily commercial species, just something to provide tree cover."

Plantations are forests, but they are much less complex than the natural forest, and are less able to buffer the harshness of the environment and climate. Usually there are only one or a few species, and often all are cloned from a single parent plant, each tree genetically identical to the others. Pests and diseases can sweep through a plantation, whereas the great variety of a natural forest is a barrier to epidemics. Erosion and runoff are greater from plantations than from natural forests, because there is less leaf litter and less secondary vegetation to absorb rainfall. There is also evidence that even-aged stands of trees are not as efficient at absorbing heavy rains as are varied natural stands. And some foresters are worried that, although in many cases plantations appear to make economic sense, we have not enough experience with industrial plantations in tropical rainforests to know whether rising costs and declining yields, due to loss of soil fertility and pest problems, will eventually undermine their attractiveness. "In the Amazon," according to Eneas Salati, of Brazil's Center for Nuclear Energy and Agriculture, "scientists have worked on this for less than twenty years, even on experimental stations, and only at Jarí has it been applied on a large scale." The owner, Daniel Ludwig, poured a billion dollars into the Jarí project, a logging operation in the Amazon Basin including 250,000 acres of plantations and a pulp and paper mill, before he abandoned it in 1982.

"We are worried about the problems that could arise from too extensive a planting of monocultures," Kanapi told me. "Fire could be a problem elsewhere, but here it rains too much. We also hope that the natural vegetation growing between the plantation areas will act as a buffer against pests." But PICOP is planting in large blocks, the biggest of which, consisting of only two species, *bagras* and falcata, will eventually cover 55,000 acres—85 square miles.

Ecologists are against too much planting of introduced (exotic) species because local wildlife depends upon native species. But some developers believe that by using species or varieties from other countries or regions, they will overcome the problems of monoculture plantations, on the principle that the exotics will have no natural enemies in their new habitat. Unfortunately, the size of the area planted to a single tree or crop is more important than whether or not it is a native plant in determining how susceptible it is to pests. And once exotics are introduced, many pests switch to them from native hosts, possibly attracted by the very fact that they are relatively pest-free. Tropical ecologist Gary Hartshorn warns that "there are numerous examples of exotic species growing beautifully in plantations for a few years, then abruptly ceasing to grow or even dying." Already PICOP has had to replant 30,000 acres of a variety of *bagras* from Papua New Guinea that was attacked by insect pests.

PICOP has had one other major plantation failure. The company planted more than twenty-five thousand acres of pines, which it hoped would supply long-fibered wood for pulp making. "That pine plantation," Executive Vice-President Rogelio Salazar told me, "looks like a waste of time and money. Our mortality rates are high, very high. If they grew like we thought they should, it would have been cheaper than importing." Now, according to Kanapi, the pines won't be harvested at all. At $400 per acre, that was a $10-million mistake.

. . .

Apart from the Manobos, a forest tribe that had long lived there, the PICOP concession was uninhabited and quite unspoiled when the company first acquired its logging rights. But it is now crisscrossed by 1,200 miles of all-weather roads that have opened the area to settlers looking for jobs and farmland—and interfering with PICOP's carefully thought out management plans.

Working with the local branch of the Bureau of Forestry Development (BFD), which is conveniently located inside the concession near the company's administrative offices, PICOP has managed to resettle about one thousand families in the Don Vicente and Doña Carmen Resettlement villages. Each family was given a prefabricated wooden house and 7.5 acres to which the government helped them obtain legal titles. "Some families were reluctant to be resettled," a PICOP official told me, "but PICOP could see that these people wouldn't use their land, so we resettled them and utilized their land." The villages are on public land that is adjacent to the concession. Each has a government school, but no electricity or piped water. The first one, Doña Carmen—named after the wife of Andrés Soriano—was established in the mid-1970s. Today many new arrivals gravitate to the towns, which are now surrounded by their own suburbs or shantytowns.

Despite the incentives of the resettlement program, a census carried out by the BFD at the end of 1981 found eleven thousand people—two thousand families—occupying fifteen thousand acres inside the PICOP concession. "Squatters are one of the biggest problems we have," said Robert Dormendo, PICOP's man in charge of forestry management and protection.

> These are the people we can't push back any more. Most are not tribal people, but most have been here since before the war. We would like to contain them so that we can manage them. We plan to organize them into communities, into cooperatives. We'll help them with the development of farms, and we would like them to plant a portion of the land where they are staying with trees. PICOP will be a ready market. We have to provide some direction that will be advantageous to the government, but one of the problems we have is to convince them.

Trying to shift the local indigenous people, the Manobo, from their land has been even more difficult for PICOP. The Manobo have lived in the forests of Mindanao for many centuries. One Manobo subgroup, the Tasaday, remained isolated, living a Stone Age life, until "discovered" in 1971 in the south of the island. The chairman of PICOP, Andrés Soriano, is on the board of PANAMIN, the government's much-criticized agency responsible for tribal people. "The Company wants tribals to move into resettlement villages, but they don't want to leave their sacred land, where their ancestors are buried," Kanapi told me. They shouldn't have to. The government has instituted something called the communal forest lease for tribal groups. Under this scheme, the land rights are given to the community, not to individuals, but the lease is only good for twenty-five years and so far it has been used in only two places in the Philippines, neither of which is in Mindanao. Nevertheless, although the Manobos do not have formal title to their traditional lands, they are allowed to remain there under certain conditions.

Land rights in the Philippines are incredibly confused. Under martial law the country was governed by a series of presidential decrees, which were often contradictory. Marcos has declared all land with more than an 18 percent slope (that's 56 percent of all the land in the Philippines) to be publicly owned forestland—whether or not there are trees on it—unless it was already privately owned. Traditional ownership does not count, although almost all the country's tribal groups live within these public forestlands. But presidential decree (PD) 410 sets out a long and complicated process through which indigenous groups can get title to their ancestral lands. The decree has never been properly implemented and no tribal group has yet obtained title this way. A pre–martial rule law "gave tacit recognition that as prior occupants, forest dwellers had rights within the forest to occupy and farm sufficient land for their basic needs." But that principle was ignored by PD 705, which defines tribal groups as "primitive tribes . . . living primitively" and lumps them together with *kaingiñeros* (peasant farmers) as squatters. Still, tribal groups—and anyone else living on public land—can stay there, according to PD 705, *if* they have tilled the land since May 19, 1975. And providing they accept any restrictions the Forestry Department imposes on their activities. Unless, that is, the land on which they are living is classed as a critical watershed or a national park or they are judged to be doing ecological damage. Recently, President Marcos extended the cutoff date to December 31, 1981, but government bureaucrats hold different opinions as to whether the extension is optional or mandatory. In any case being allowed to stay on the land does not give one a formal title to it.

There are various kinds of title, and various ways to acquire one. The Bureau of Forestry Development may give farmers a Forestry Occupancy Permit (FOP), good for two years, providing they follow BFD guidelines about planting trees with their crops, as a safeguard against erosion. Some families are required to work in government reforestation projects. Others may be forcibly resettled. Some ten thousand such permits have been issued, and the government says that the program has drastically slowed the rate of damage from slash-and-burn agriculture.

The FOP has now been succeeded by the Stewardship Contract, which lasts for twenty-five years and can be used as collateral, although the land itself cannot be sold. Stewardship Contracts, with the longer period of tenure, are meant to give poor families a stronger sense of security and, the authorities hope, a stronger motivation to fight environmental degradation. The longer tenure, however, is not viewed with universal approval. "PICOP is requesting us not to issue Stewardship Contracts," Mariano Valera, head of the BFD's Timber Management Division, said. "It has to be discussed with PICOP because they'll have to be involved if we want harmony." PICOP rests its case on the fact that President Marcos's Letter of Instruction Number 1260 guarantees the company exclusive use of its concession.

The main idea, Valera told me, is to "allow these people to occupy the lands they live on, and yet do not release it. . . . In the rural areas the people are very poor. If you pay them well enough, they will be happy to plant any kind of tree; it doesn't really matter what kind. You keep your fingers there, saying, 'That is not the way to do it. Do it this way.' "

Manobos and settlers who want to stay in the PICOP concession can't legally be forced to move—if they have been on the land since December 31, 1981 (or is it May 19, 1975?). But, Aida Bruzon, an official of the local branch of the BFD, says, "Their area has been surveyed. They can't extend or encroach on the forest. If they do, they'll be subject to prosecution. We have prosecuted and are prosecuting even tribals for this, after a warning." The Anti-Slavery Society, a London-based human rights group active since Victorian times, bitterly criticizes this vigilance, noting that "companies cutting up to 3,000 trees per day are to be tolerated and encouraged, but whole communities that cut less than this amount in one year are to be restrained and even forcibly removed."

The efficiency of PICOP's own BFD branch—due partly to adequate resources and funding—is in sharp contrast with other areas of the Philippines, where slash-and-burn agriculture is a much more serious problem. I spoke to Sesenio Sereño, the head of the Watershed Protection Division in one of the BFD's twelve regional offices. His region, Central Visayas, has only 7 percent forest cover, the lowest in the country. There are an estimated 22,000 families farming on watersheds in Central Visayas, covering an area of about 125,000 acres. In 1980 his department was set the goal of helping 1,900 slash-and-burn families (more than 10,000 people) to achieve a stable agriculture on the slopes where they were living or to resettle elsewhere. The budget for this was less than $100,000. "In 1981 I got the same amount of money and an obligation to help another 1,900 families, but should I leave the first 1,900 behind? We have to continue to work with them or we'll never achieve our objective." Perhaps luckily for his peace of mind, Sereño doesn't know exactly how bad things are in his region. "I cannot even guess how much watershed I am responsible for. I have requested funds for an inventory. I have no assistance, although sometimes the soil technologist helps me process permits."

BFD officials in the PICOP concession say that their squatter problems have diminished lately. "Recently," Aida Bruzon said, "we have had no problem because of the presence of the army. San Vicente is now a military area. Troops are deployed in barrios, cities, municipalities—all over. With their presence the people were really afraid to go into the forest. They are a great help to the bureau. We coordinate with them."

According to Bishop Carmelo Morelos, the Roman Catholic bishop of Butuan in the neighboring province to PICOP, "the military have a high presence where there are big investments, especially foreign investments. If there is trouble, they send in the military and concomitant with that is discontent, abuses, and so forth, and that sends some people over to the other side. Then they send in more military." "The other side" is how Filipinos refer to the anti-Marcos New People's Army (NPA). The military is here to prevent the other side from gaining control of the area. The vast majority of peasants want simply to stay out of the conflict, but as one told me, "The other side comes here, asking for food. They're just ordinary Filipinos like us. Why should we refuse them?"

. . .

PICOP's tree-farming scheme has impressed development planners, foresters, and ecologists alike. The company encourages nearby farmers to plant falcata by offering technical assistance, guaranteeing to buy the trees, and helping them to borrow money to finance the expansion. The scheme is credited with increasing PICOP's supply of wood, reclaiming degraded land with an ecologically appropriate crop, and providing poor farmers with a good living on their own land so that they do not encroach on the concession. "There may not be another instance in the world where slash and burn is discouraged more effectively," wrote the editor of *World Wood* in 1981. There are now more than four thousand farmers growing trees for PICOP. Their plantations cover almost seventy-five thousand acres and supply one third of all the wood the company uses.

Edmundo Losara came to this area in 1960 from the western part of the island. He is married, with three very shy children, aged two, four, and seven. At first he was a squatter and had some land in the concession. Then in 1964 he moved one and a half miles away to Doña Carmen and a few years later got work as a mechanic at PICOP. When Marcos declared martial law in 1972, he also decreed that unused land could be confiscated. So Losara quit his job at PICOP and went back to his farm. In 1976 the company and the government helped him get title to his land. He is now a successful tree farmer with ten acres of falcata, which he grows in combination with coffee and cocoa. Losara has a water buffalo to help with his farming. He did not need a loan to establish his plantation. His wooden house has no running water or electricity, but it is surrounded by a charming English-cottage-type garden, with a tiny patch of lawn and bougainvilleas, palms, and crotons in place of roses, hollyhocks, and privet.

PICOP started the tree-farming program on its own, inspired by the experience of its partner, International Paper Company, with a similar scheme in the southeastern United States. But the project really took off in 1974, when the Development Bank of the Philippines, financed by the World Bank, set aside $2 million for loans to participants.

Anyone with at least 12 acres of land within a sixty-mile radius of the PICOP mill can take part in the scheme, and more than three hundred PICOP employees, including some high-ranking ones, do. I asked Ricardo Santiago, the company's vice-president for administration, and his wife about this when they took me to lunch at the clubhouse the company maintains for senior staff. All the PICOP employees who had tree farms, they said, availed themselves of the World Bank loans. Most of their farms were from 50 to 75 acres, though one man, who had recently retired, had 2,500 acres. The Santiagos made a good joke out of the fact that they hadn't got a tree farm. They missed out on buying land when it was "going for centavos. Every year we thought we were leaving—and we're still here fifteen years later!"

Landless people are not eligible for the project. Almost all the participants already owned the land for their tree farms and they pay 12 percent interest on the tree-farming loans. A few have legal proof of occupancy only; they pay 14 percent. According to the International Agro-Forestry Development Corporation in Mindanao, it costs more than $2,500 to start a twelve-acre tree farm and maintain it for five years. The loans available are inadequate, though they are intended to cover 75 percent of the costs of establishing and maintaining, but not harvesting, the plantation. The DBP lends tree farmers less than $100 per acre. The interest costs of the loan eat up most of the income from the first harvest and would leave only one sixth of the gross income expected from two rotations of the trees, a sixteen-year period. During the eight years that it takes the trees to come to maturity, the farmers don't have to make repayments. When the trees mature, the company buys them, first deducting the loan repayments. Under the circumstances it is not surprising that 70 percent of the growers don't borrow money to start their farms.

The average tree farm is twenty-five acres and the average income of those taking part in the scheme is $925, three times the average income in the Philippines. So the program reaches not poor and landless squatters but a fairly well off group of people, landowners with an above-average income. Similarly, only one quarter of the land converted to tree farming had previously been idle or degraded. The rest had been used to grow food or other tree crops. Thus the tree-farming scheme has failed in two of its three aims—helping poor people and reclaiming wastelands. Nonetheless, from the company's point of view, the project is a success, providing a steady supply of pulpwood at low overhead and labor costs. And it has strengthened PICOP's position in the region, discouraging disruption by the "other side." "Even the NPA doesn't want to disturb operations at PICOP, since they know that it's a source of income for the people," Salazar told me.

For the tree farmers, however, the project has some serious problems. Only half of those interviewed in a 1981 survey said that they would continue in the program. One of their main complaints is that the cost of harvesting is not included in the project loan. Since they don't get paid until they deliver the trees to PICOP, many of the smallholders don't have the cash to pay for help in harvesting and transporting the trees. As a result, they are forced to hire contractors who offer credit but take as much as 45 percent of the final price. One reason that no one foresaw this problem is that the farmers did not adhere to the company's original plan.

That plan was that smallholders would plant 80 percent of their land with falcata over a four-year period (keeping the rest for food crops), apply fertilizer twice a year, weed three times a year in the first two years, and, once the first trees matured, harvest one eighth of the plantation each year for eight years. In return the company was to provide seedlings for the farmers to buy and give them advice. PICOP did this well; it hired one technical advisor for every 173 tree farmers, organized tree farmer associations, and broadcast advice on a daily radio program.

But most farmers found it easier and cheaper to plant their entire area at once. Naturally this meant that eight years later the whole plantation was ready to be harvested at the same time. PICOP therefore found itself one year with much more falcata than it expected and much less in subsequent years. And farmers had more wood than they could handle and harvest by themselves. Farmers also, despite PICOP's strong technical support, used less fertilizer and weeded less often than the company recommended. However, that savings of materials and labor improved the farmer's profit on the trees, a fact that was especially important in view of the generally low prices PICOP was able to pay (due, the company says, to government price controls on the newsprint produced from the pulpwood). Buyers in Japan, Korea, and Taiwan pay three times as much for falcata, but, although the PICOP farmers are free to sell elsewhere if they notify the company in advance, they do not do so for logistical reasons.

In 1982 there was a freak typhoon in the region of the PICOP concession. It caused more than $20 million worth of damage to the plantations of PICOP and the tree farmers. A lot of the damaged wood was salvageable—more than PICOP could handle. Because the company wanted to recoup its own losses first, it decided to accept only a small amount of wood from its tree farmers. As a result many tree farmers were unable to sell the crop that had occupied their time, money, and land for several years and faced losing their land because they could not keep up their loan repayments. The incident caused much bitterness among local people. This and the discouragement of many of the tree farmers is bad news for PICOP, which still wants to expand the program.

. . .

Unlike many, probably most, tropical logging companies, PICOP cares about its reputation locally because it plans to keep working in the same area for many years. According to a senior official at the Asian Development Bank, the typical attitude of tropical rainforest loggers is "if you can't get a return in two or three years, you're not even going to bother. This is not only true of foreign investors. In the Philippines, for example, it's particularly true of Filipinos. They don't trust the future. Everyone wants to get their money out of the country. They'd rather invest it in California." Scott Jackson, a manager of Georgia-Pacific's concession in eastern Kalimantan (Indonesian Borneo), told me that the average plywood mill in Indonesia recouped its costs within six to eight years, sometimes in only two and a half years. "They want a fast pay-back because they don't know what will happen in the government or whether the forests will last."

PICOP has invested enormous sums of money, not only in equipment, but also research. Its engineers developed a revolutionary way to make pulp and paper from tropical trees, which have short fibers rather than the long fibers of traditional pulpwood. The company's forestry staff of fifty-two, half of whom are devoted to research, "would make the forestry agencies of most developing countries envious," according to the timber trade journal *World Wood.* They compare ecological effects and relative costs of, for example, different ways of extracting logs. Researchers at the company's four nurseries work on improving techniques of transplanting and maintaining seedlings and breeding better varieties.

But PICOP's technical successes are not reflected on its balance sheet. In 1980 it lost $20 million. In 1981 it lost $35 million. In 1982 it lost slightly less, about $30 million. In fact, the company actually made an operating profit in all those years, but it was more than wiped out by the cost of paying interest on its huge borrowings. PICOP's financial problems started in 1970, when, as it was building the short-fibered newsprint mill, the peso was devalued by 40 percent. Then followed a series of catastrophes, from the 1973 oil crisis to the government's failure to connect the company to the national power grid in 1974, as promised. (The company now hopes to be connected in 1984.) PICOP was forced to borrow hugely, mostly from American banks. At the end of 1981, the Philippine government rescued the company from defaulting, and, along with the Development Bank of the Philippines, took 46 percent of the company, leaving the Soriano Corporation and International Paper Company with the rest.

Yet the company also enjoys a number of privileges. Its concession is twice as big as allowed under Philippine law. It pays no royalties on its paper production. It is exempt from sales tax, import tax, and forestry taxes. It is one of the handful of companies in the Philippines allowed to export logs. Salazar, who was brought in in 1979 from another Soriano company as a tough guy to tighten PICOP's belt, is positive about the future. "We started badly, with technical problems. Now that we have the technology, we think we can make it."

The first step for Salazar was laying off almost half the company's workforce of fifteen thousand. The company acknowledges the devastating effect its layoffs have had on the local population. Even those not employed by the company depend indirectly upon the money PICOP workers inject into the local economy. In an attempt to reduce unemployment in the area, PICOP's management hired buses to carry jobless volunteers back to whatever part of the country they originally came from.

Even at the best of times there is a huge gulf between the laborers and the managerial staff, who live in Bay View Hills, the residential section of the fenced-off PICOP compound. Their ranch-style houses set in large, well-manicured gardens contrast sharply with the shacks of Mangagoy, the nearby shantytown where most of the laborers live. Mangagoy is a company town; 80 percent of its forty thousand residents came here from elsewhere to work for PICOP. Residents of Bay View Hills get free housing, electricity, and water. Not Mangagoy. And Rogelio Salazar, for one, is tired of being accused of unfair treatment. "When we started the town, we supplied it with light, water, and so forth. We encouraged businessmen to form their own enterprises as contractors. We asked them to go into tree farming. We got all these things started, but we felt they should evolve into a self-sufficient entity. The government hasn't spent any money on that place. They expect PICOP to do it."

So did a priest I spoke to, who pointed out that while the PICOP compound is supplied with clean water, the town's supply is not fit to drink. All employees are entitled to medical treatment at the company's hospital. Staff members and their families get private rooms. Laborers are put in wards. "Medicines are free if they are available, but they are not often available," said the priest. It is "a trick of the company," he said, to lay people off and rehire them—keeping them as casual workers. Only permanent employees are entitled to hospitalization and other benefits. A PICOP worker makes about 600 pesos ($75) a month, which is not, according to the priest, a fair living wage. "Still," he said, "PICOP is one of the best companies in the Philippines in terms of treatment of workers."

11
Pioneers

When Alfred Wallace visited Bali in 1856, on the journey that helped shape his views on evolution, he was struck by its "impressive system of irrigation that would be the pride of the best cultivated parts of Europe." That system, along with the island's fertile soils—laid down over thousands of years by erupting volcanoes—and its social cohesion, enables Bali to support a large population with the most intensive agriculture in the world. But as Bali's population has grown, the subdivisions of a family's lands among its sons have become smaller and smaller. Today the average farming family on Bali has only one quarter of an acre. Many younger sons have no hope of ever owning any land. Since 1960, 120,000 Balinese have left what they call Paradise Island, hoping to find land and a better life elsewhere.

In 1980 Ketutgede Suwendri and his wife, both in their early thirties, left Bali, left their families and their two children behind, and came to southern Sumatra looking for land. They chose the village of Karangsari in Lampung Province because they had friends there.

Twenty years ago there were no people in Karangsari. There was no village, only a forest called Way Kambas, so rich in its animal life that in 1937 the Dutch declared it a game reserve. Tapirs, elephants, tigers, macaques, leaf monkeys, gibbons, one-horned rhinos, barking deer, and Malayan sunbears then roamed through the 300,000-acre reserve. Several types of crocodile lived in the swampy coastal forests. In the 1930s there were so many snakes that in one day's walking along the Penet River, a Dutch colonial officer named Pieters killed ten pythons—although he shot at only a fraction of the number he saw. The forest was dominated by majestic dipterocarps. It is impossible to say what other plants then grew there, because in 1968, before botanists had had a chance to study it, loggers moved into the reserve. They had government licenses to fell trees over a 115,000-acre territory, but they harvested more than twice that area. Eleven years later, when they finally departed, only one tenth of the forest was left relatively undamaged. Seventy percent had been converted, by exposure to sun, brushfires, and the weight of the heavy machines, to scrubby brush or elephant grass—a coarse weed, known locally as *alang alang,* that invades degraded soils. Only in a few swampy patches along the coast, where heavy logging equipment couldn't operate, did the forest survive.

Way Kambas is a triangle on the east side of Sumatra with its longest side on the coast of the Java Sea. It includes several types of forest, ranging from coastal mangroves to freshwater swamp forest to dipterocarp forest, each with its own characteristic plant and animal life. Although there has never been a detailed inventory of the plant life of Way Kambas, it is safe to say that before the loggers appeared, its forests contained several thousand plant species. Some —no one will ever know how many—have gone forever.

When the Suwendris arrived, most of the damage to Way Kambas had already been done. They did not clear forest for their home and farm because there was so little forest left. Instead, they settled on the desolate grassland that had largely replaced the trees. They spent six months burning and clearing eight acres to make it ready for planting. Now they grow maize, cassava, bananas, peanuts, coconuts, taro, and, in the rainy season, for there is no irrigation, rice. Ducks and chickens scamper about the garden. The house they built has woven grass walls and a deep sloping thatched roof that creates a shaded veranda.

A cage with two small birds hangs in the relative cool of the veranda. Over the entrance to the house is a small sculpture of a deer carved from a coconut shell. The dirt floor is bare and swept clean.

In the cooking area pots and pans are neatly stacked on shelves; on the floor a cooking fire is laid, ready to light. At the other end of the house a double bed on stilts is draped with a batik print and veiled by a mosquito net. A second, smaller bed, also raised high off the floor, is covered with grass matting. A huge radio, of the sort that disturbs the peace in Central Park, is on a shelf nearby. In the northeast corner of the house, corresponding to the correct position for a temple in a Balinese family compound, hangs a small model of a Hindu temple, made of fabric and held in shape by bamboo stiffeners. As far as they are able in this foreign and somewhat hostile place the Suwendris have re-created the home they left behind in Bali.

"It's not yet as good here as Bali," Mr. Suwendri said, laughing, "but I won't go back there. We left because there was no land for us there, so it's very hard to make a living. Here we have land, eight acres. Plants grow easily, especially cassava, which we sell to the factory. We have friends in Karangsari who have been here for twenty years. They have built temples, and there is Balinese dancing at festival times."

In 1937, when Way Kambas was designated a game reserve, there were a few small villages on the outskirts: along the borders and on the coast. They were abandoned during the war or earlier. In the 1930s, for example, the inhabitants of Puto village deserted it to escape from man-eating tigers. In 1963 the Forestry Department helped newcomers establish Karangsari and two other villages inside the boundaries of the game reserve. Five years later the department gave a number of logging companies licenses to fell timber inside Way Kambas. The roads they built opened the area to settlers. There are now about eighteen thousand people living in fifteen villages in the reserve. Six of these are farming villages like Karangsari. Their homes and fields take up about twelve thousand acres at the southern border of the reserve. The rest are small, and for the most part seasonal, fishing settlements strung out along the coast. In 1969 the provincial governor gave Karangsari legal status as a village. The provincial government built schools in the Way Kambas villages and continued to encourage them to expand.

Today 566 families live in Karangsari. It looks like a typical Balinese village, except that whereas Balinese houses (or rather family compounds, for each family has separate buildings for eating, sleeping, cooking, and praying) nestle together, here everything is spaced out to take advantage of the land available. Though the village here is more spread out, the houses have the same thatched or tile roofs, the same beautifully elaborate stone carvings, the same strictly ordered arrangement of domestic buildings and temples. Not only did the pioneers of Karangsari struggle to make this unfamiliar land productive, they worked to make it home. Flowering shrubs and vines—hibiscus and bougainvillea are common—surround every house. The roadsides are lined with hedges and dotted with trees.

It is impossible for the villagers to farm in precisely the same way here in Karangsari as they learned to do in Bali. For one thing, the soil is different. It is older, and, because its store of nutrients has been depleted over the years, less fertile—although the area was blessed by the explosion in 1883 of Krakatau off the southern tip of Sumatra, seventy miles away. Krakatau deposited a layer of fertile dust, up to two inches deep in places, over the southern part of Way Kambas. But the soil has been badly affected by the destruction of the forest. Removing the trees exposed the soil to the unrelenting glare of the sun, and the heavy logging machines compacted the soil and damaged its ability to retain water. Because Karangsari is not irrigated, the people can grow rice here only in the rainy season. Cassava, which requires little water and which a nearby factory processes into tapioca, is the villagers' main crop. But the farmers of Karangsari have retained at least one important feature of Balinese agriculture, the idea of combining many different crops in a small area. One villager I spoke to grew maize, cassava, bananas, peanuts, coconuts, taro, rice, peppers, cloves, fruit trees, including rambutan and mango, as well as various local herbs and flowers for cooking, religious and ceremonial uses, and just for pleasure. He also kept chickens, ducks, and pigs, as well as an ox to help with the plowing.

Life is not easy in Karangsari, but over the years the villagers have created a social and an agricultural oasis. They have good relations with their neighbors, most of whom have emigrated from Java. None of the people I spoke to wanted to return to Bali, though they missed their families and friends there. "Everybody is more happy here," the schoolteacher, Wayan Gatar, said. "In Bali we haven't fields. Here we get good crops. We're far from the cities so things are cheaper and we have no fear of crime."

Bali and Java are the most fertile and crowded of the thirteen-thousand-island Indonesian archipelago. In parts of rural Java where the land is irrigated, as many as five thousand people are crowded onto one square mile. Eighty percent of Java's ninety-two million people live in the countryside; of these 40 percent have no land at all and another 35 percent have not enough land to grow the crops they need to feed their families. One cause of the crowding is unfair land distribution. One third of the land in Java is in the hands of 1 percent of the landowners, despite an ineffectual attempt at land reform in the 1960s.

The so-called outer islands—notably Sumatra, Kalimantan (the Indonesian section of Borneo), Sulawesi, and Irian Jaya (the western half of the island of New Guinea)—are not so hospitable to man, and man has by and large shunned them. Their soils are generally older, more weathered, and relatively infertile, not suitable for intensive agriculture or dense populations. For the most part, these islands have until recently been inhabited only by small isolated groups of people who hunt, fish, gather, and practice shifting cultivation in the forest. Kalimantan has fewer than twenty-six people per square mile; Irian Jaya fewer than eight. Thanks to these long-undisturbed, forested islands, Indonesia has one tenth of the world's tropical rainforest; only Brazil has more. To the world at large, the outer islands of Indonesia have for many centuries been unknown wild places, dangerous and enticing frontiers. But by the beginning of the twentieth century, as the inner islands grew increasingly crowded, those empty and unknown lands began to seem a wasted resource. Resettlement seemed the best way to "use" them.

The Transmigration Program, as it is known, is the most ambitious colonization scheme in the world. In the 1950s President Sukarno called transmigration "a matter of life and death for the Indonesian nation." Indonesia, he said, did not need birth control. Instead Java's population was to be reduced from 54 million to 31 million between 1950 and 1985 by moving 140 million people to the outer islands. Plans for the program changed frequently as the government failed to meet the ambitious goals it had set itself. The first Five-Year Plan (1956–1960) aimed to move 2 million people but managed to move only 127,000. The target was to be lowered for the 1961–1968 Eight-Year Plan to 125,000 people a year. At Sukarno's insistence, the figure was raised to 1.5 million people a year, equal to the annual increase in the population of Java. But between 1961 and 1965, when Sukarno lost power, only 135,000 people had been moved. In 1966, when Suharto took control of the government, his cabinet set a goal of moving 2 million people a year. In fact, in the eight years to 1974, fewer than 200,000 people were moved. The current Five-Year Plan calls for 2.5 million people to move from the most crowded islands to the so-called outer islands between 1979 and 1984. It is a daunting goal, but in the first four years the government moved an unprecedented 1.5 million. Encouraged by that success, the government wants to move twice as many in the next five years.

Transmigration has already had an enormous impact on the country's rainforests. Parts of Sumatra and Sulawesi, formerly heavily forested and virtually uninhabited, are now as crowded and as devoid of trees as parts of Java. The ecological and social costs of replacing the forests with large numbers of poor and inexperienced farmers have been high. Settlers have clashed with local people, who resent the invasion of their lands. Wild animals, including elephants and tigers, have lost so much of their range that they have been forced to prey upon the communities that have displaced them. Indiscriminate land clearing has damaged the soil, eroded the land, silted reservoirs, disrupted irrigation, and caused serious flooding.

. . .

Although there is some good agricultural land in the outer islands, most of it is already spoken for—by indigenous people, spontaneous migrants, and owners of large plantations. An unpublished survey of the Transmigration Program by the World Bank found that "the land available for transmigration increasingly requires the more difficult establishment of the settlements on uplands or tidal lands." Customary law in Indonesia, called *adat,* does not provide for individual ownership of land. Instead, the local community, the *marga,* owns land collectively. The government has to negotiate with representatives of the *marga* in order to obtain land for transmigration. The same World Bank report noted that "as these leaders often tend to be more knowledgeable of the quality of their land than the negotiating Government of Indonesia representatives, the best land is generally not secured for transmigrants."

Most of the settlements have so far been in forested areas. Although clearing land with machetes and chain saws is less damaging to the soil than using heavy equipment, the rush to meet the target of half a million families in five years means that most land is cleared by big machines. Generally the trees are knocked down by bulldozers and then burned—often even before the commercially valuable species have been taken out. Heavy clearing equipment compacts the soil so that it loses much of its capacity for retaining water. Hard rains can then erode the land, silt up nearby rivers and dams, and lead to flash flooding as the waters sheet along the surface of the ground. The wheels and tracks of the clearing machines scrape off much of the thin layer of topsoil, exposing subsoils that are less capable of receiving and storing nutrients, so that fertilizers become less effective.

In 1979 President Suharto issued a presidential decree prohibiting the clear-cutting of primary forests for transmigration. The plan was to settle families on Indonesia's vast and expanding area of *alang alang* grassland. The problem with Indonesia's eighty thousand square miles of *alang alang* grassland, and the reason the government has not been able to settle many transmigrants on it, is that most of it is already owned by someone. This is true by definition, since *alang alang* grassland is almost always former forest that was cut and repeatedly burned to fertilize the soil with ash and control weedy secondary growth. "*Adat* says that working the land brings it into ownership, even if that land is later abandoned," said Soeriaatmadja, an affable ecologist and advisor to Indonesia's environment minister, Emil Salim. "So how can we settle people there? How could we compensate the owners under *adat* law unless we pass a law that says that if you cultivate land and destroy it, turn it into *alang alang* grassland, then the government will take it over to rehabilitate it? This might make sense from an environmental point of view, but I doubt that we could have such a law. There is a question of human rights here."

Two thirds of government-sponsored transmigrants are landless peasants, the poorest of the countryside; another 10 percent are homeless city dwellers. Many are not experienced farmers. A 1976 study showed that 45 percent had never before grown rice, the staple crop of Indonesia and of most transmigration settlements. Spontaneous transmigrants are among the most successful settlers, both because they are highly motivated enough to move on their own initiative and because they can choose where to live—usually near friends who can help them through the difficult early years. The same survey in 1976 showed that seven years after settling, most spontaneous migrants have a higher standard of living than "official" transmigrants. Recognizing this, the government of Indonesia is now trying to encourage the spontaneous migrants by reimbursing some of the costs of moving and in some cases providing them with the same health and education services the official transmigrants get.

The least successful—and the least popular with other settlers and local neighbors—are the urban poor. "It has been a bad experience sending people recruited from the slums in Jakarta," I was told by Soedjino, an assistant junior minister of transmigration. (Indonesian names, one Indonesian told me, are so complicated that many use just one in their working lives, especially with foreigners.) "They won't cultivate the land. They go back to town. That's our biggest failure. That's why many regions reject transmigrants from Jakarta." Up until 1973 people from Jakarta were often almost press-ganged into moving. Now volunteers outnumber available places on the program, government officials say. But one American working as a consultant on the project told me that as officials strained to meet the new targets, there was a danger of their becoming less selective, both about proposed sites and about whom they move. "It's still a voluntary program, but they are increasing pressure on people to sign up. The government is upping the hype, but now people know more about the program than they used to, so they're less easily fooled."

. . .

Few migrants go back to their villages, but stories about transmigration trickle back, not all of them encouraging. Each family is promised a house and from five to eight acres, about half of which should be cleared before they arrive. But the logistical problems of building whole towns in empty and isolated areas plague the program. Sometimes the government department responsible for moving the settlers to their new home gets them there before other departments have finished clearing the land, building roads and houses, or ensuring a water supply. This is common enough that the government now sometimes offers the people a small fee to build their own houses. According to a survey of the program carried out by the International Development Research Center (IDRC) in Canada, "studies of transmigration programmes frequently cited cases where contractors did not finish their tasks before the settlers arrived, houses were not built, roads were not constructed, implements were not available, rations were insufficient, and basic medical assistance was not provided."

Transmigrants all receive food rations for the first year or two. The standard issue is forty pounds of rice per month for the man, twenty-two for his wife, and sixteen apiece for three children. Each family also receives a supply of salt, salted fish, cooking oil, kerosene, sugar, and washing soap each month. In theory a husband and wife can bring only three children with them, but studies have found that one third of the transmigrant families number six or more when they move. Extra children do not entitle a family to extra food. Settlers still need cash, and that they can only get by selling their surplus crops, if they have any. But, as the IDRC report points out, "Even if he enjoys a bumper crop, he might find it extremely difficult to sell his produce since transportation to the nearest town is difficult and expensive."

Most settlements are too small to support a diverse community, with craftsmen providing services and paying for farmers' produce. "You need at least ten thousand families to have social and economic diversity," says Gloria Davis, the official who oversees the World Bank's loans to the project. "At that level maybe one third of the population could be in something other than agriculture, providing services. But only if the farming families generate a large enough surplus."

Settlement sites are often too small and far from one another to justify the expense of providing local services, such as post offices, schools, and clinics. In 1977–78 the Indonesian government built a total of three medical clinics, seven elementary schools, and five houses of worship on transmigration sites. Perhaps luckily, in the same year officials were able to move fewer than 60 percent of the 105,000 people they had planned to move. Rachmat Wiradisuria, assistant minister of the environment, told me that "not being able to meet our target figures for transmigration was a blessing in disguise. It gave us a chance to learn from our mistakes."

Officials at the World Bank believe that the biggest weakness in the Transmigration Program is the way sites for new settlements are chosen. The government has generally overrated the quality of the land available. Frequently it has been forced by failure to tone down its goals. In 1967 it announced a plan to turn virtually all the country's tidal swamp forests, 13 million acres, into rice-growing areas by 1982. Two years later the figure was cut to 1.25 million acres. Problems were so great in one tidal swamp settlement, Barambai in Kalimantan, according to a study of the area, that 16 percent of the families abandoned their sites within the first three years. In 1974 there was a new plan to convert 2.5 million acres of tidal swamp to agriculture, but that was later reduced to 600,000 acres. For this reason, the World Bank has earmarked $120 million of the $350 million it has lent to the program to be spent on developing better ways of selecting new sites. A bank official warns that better planning may bring "a big conflict over site availability. Already there is a lot of tension in the system because the consultants are recommending exclusion of more sites than under plan-as-you-proceed. I believe the government will run out of sites before they meet their resettlement goals, but the government doesn't share that view. One problem is that appearances are deceptive. It looks like there's a lot of land in East Kalimantan and Irian Jaya, but it's impractical to settle there."

The procedure laid down by the government for establishing transmigrant communities calls for five years of planning and preparation before anyone is moved to the new site, and another three years of advice and assistance before the community is on its own. "If we had enough time to go with this system, we could eliminate many of the dangers," said Soedjino. "Unfortunately, sometimes we have to finish in three years, or sometimes two years. Sometimes we plan and act simultaneously, what we call here plan-as-you-proceed." One high official of a foreign funding agency described plan-as-you-proceed thus: "As soon as anyone finds any land that is vaguely empty, you move people in."

"Now it's getting difficult to get enough land to accommodate the target," agreed Soedjino. With the flatlands fully occupied in most parts of the country, more and more new settlements are on hilly land. Because clearing such land for annual crops has led to severe erosion on many sites, the World Bank is eager to promote tree crops, although tree-crop settlements, because they require expensive fertilizers, cost twice as much per family as standard, low-input settlements.

. . .

I asked the office of the governor of Bali for the names of some villages from which people were emigrating under the government program. His assistant couldn't have been more charming or helpful, but any notion I had of making my own way to one of the villages they named and seeing for myself what sort of conditions transmigrants were leaving behind was quickly quashed—killed by kindness. Before I knew it, an official delegation, including the personal assistant to the governor of Bali and the district chief, was escorting me to Petulu, a village from which about twelve families, one quarter of the population, were soon to migrate to Sulawesi.

As we drove along, I chatted with Paramartha, the governor's assistant. He is Balinese, a relaxed, cheerful, and open man, who was clearly somewhat bemused by my interest in transmigration. Paramartha's regular duties included dealing with foreigners, and whenever the conversation flagged, he had a good stock of one-liners. A passing flock of geese was dubbed "Japanese tourists"; at the end of a discussion of Balinese wood carvings, looking at the potholes we were bouncing over, he sighed and said, "Everything in Bali is so beautifully carved, even the roads." Even so, I believed his frequent comments along the lines of "Bali is not far from Paradise," since it certainly struck me that way. Although, or perhaps because, his job was to deal with visitors, he told me, "We don't want to be the second Hawaii. Hawaii has lost all its culture." Bali is an amazing holdout from the general worldwide rush toward Westernization and cultural blandness. Although hopelessly American in many ways myself, I felt positively grateful to hear locals spurning the American way. Even minor rebellions—the administrative order that keeps tourist development to a peninsula on the southern tip of the island and forbids hotels to rise higher than a palm tree, for example—seem good omens.

Paramartha's father is a bank manager, a member of the royal caste, "but my parents dropped that. They believe that if you treat people with respect and behave well, then people will treat you also with respect. It has nothing to do with titles." He has a degree in business administration and was to have studied international law in Perth, Australia. "I wanted to understand the Western attitude to life, with material and spiritual and natural things all separated. But I didn't want to lose the Balinese view of things. That's why I chose Perth, instead of Sydney or another big city." But he bowed to his parents' wish that he remain in Bali.

As we drove through one village, I noticed an open-walled schoolroom on the other side of the road. I was taken aback to see how old most of the people in the class were; then I realized it was probably an adult literacy class. The car stopped and I was helped out and led across the road. Suddenly it hit me that this was Petulu, and all those grown-ups in the schoolroom had been summoned to submit to my questions. I was horrified; all I could think of was how discourteous it was that these people, my elders (I am not consoled by my retrospective realization that many were probably in their twenties and younger than me), had been ordered to appear for my convenience. I also immediately saw that this was doomed to be a meaningless, formal occasion and that no one, probably including me, would speak freely. As I mounted the steps, accompanied by my battery of officials in suits, I felt as nervous as if I were about to appear in a play in front of a hostile audience, which, apart from the fact that the villagers were excruciatingly polite, was an appropriate response.

The head of the village welcomed us, and my delegation took its place at a long, U-shaped table, facing the hapless future transmigrants ranked meekly on benches. My mind was racing. What could I do to rescue this disastrous situation? Guilt mingled with a desire to get some useful information about transmigration. The only thing, I decided, was to make a fool of myself. I clowned with the children, winked at them, curtsied, and made them and their parents laugh—trying hard to dissociate myself from the serious-looking officials I was with. I tried to ask questions that could be answered with a shake of the head, yes or no; it's a good way to check on the translator—and mine turned out to need a fair amount of checking. Most of the Petulu transmigrants were farmers; some were wood or stone carvers. Quite a few of the men were younger sons. Some would have no land if they stayed in Petulu. No one from the village had transmigrated before, and none of the people who were about to leave had relatives or friends who had emigrated. None had seen the place they were going to. They seemed optimistic that life in Sulawesi would be better. They were praying to their ancestor spirits, asking them to come to the new place. They told me that there would be roads and wooden houses ready for them when they reached Sulawesi. The government had also promised a primary school and a health clinic, though they were less certain about electricity. The people would build their own temples. The most important thing, though, was land; Petulu had good farmland, with beautiful rice terraces, but it was overcrowded.

Two days later I was walking through a village near Petulu with Eddie Ayensu and Gerardo Budowski, both in Bali for a meeting of ecologists, botanists, and zoologists on how to give real protection to national parks and other nature reserves. As we looked at a tiny garden behind one house, which was crammed with more vegetables and fruits than even those two tropical botanists could identify, I told them about my encounter with the transmigrants and about their willingness to risk everything for the chance of a better life elsewhere. Budowski said, "Yes, they are enthusiastic; they have known only this fertile land. Send them to Sulawesi and see how their enthusiasm will change in two years."

. . .

Between 1932 and 1974, transmigration saved Java a population increase of almost one million people. But in the same period Java's population actually grew by almost forty million. With Java's population still increasing by about two million a year, government officials no longer claim that transmigration will relieve population pressure. In any case, although Java's birthrate has declined from a 3 percent annual growth to 1.7 percent, the official line in Indonesia is still that a large national population is not a problem. A current government brochure says, "The high number of population is not regarded as a burden to development; on the contrary, if guided and mobilized efficiently, it could become an effective labor force, representing a valuable asset in development and beneficial to development efforts in all fields."

If transmigration is not solving the population problem, what is the point of this massively expensive and ecologically dangerous scheme? Government publications list seven official goals of the transmigration program:

1. Improvement of the living standard
2. Regional development
3. Even population distribution
4. Equal development throughout Indonesia
5. Utilization of natural resources and human labor
6. Integrity and unity of the nation
7. Strengthening of defense and national security.

The purpose of transmigration now, Soedjino told me, is to "set up a new society better than the original one. The living conditions of the transmigrants should be better than before." Transmigration, he said, encourages regional development and adds to Indonesia's food production. I asked what contribution the transmigrants have made to the country's food supply. "Ah," he replied, "that's a good question. So far there is no study actually." In fact a Ministry of Agriculture study showed that rice from transmigration areas accounted for less than one quarter of 1 percent of the national harvest.

An economist who studied the program for the World Bank found that production was abysmally low except in the 10 percent of settlements that had tree crops or irrigated rice fields. "None of the other settlements had returns anything like what the World Bank said would be the minimum feasible in its economic justification for the program. There is a small net increase in food production, but it doesn't justify the investment. It would make much more sense to invest the money properly and buy in that small amount of rice."

People stay on transmigration sites because, by and large, they are better off there. But this is not the reason that the government is spending from $6,000 to $12,000 to settle each family. The real motive for the program, many believe, is "Javanization." The indigenous population of many of the islands under Indonesian rule (notably Kalimantan, Irian Jaya, and Timor) have rebellious tendencies. They have their own cultures, languages, and religions and do not want to be part of Indonesia. Many people on these islands bitterly deprecate transmigration as a move to swamp them with people who share an ethnic and religious identity with Indonesia's military rulers.

Even where the intention of the program is not to subordinate the native community, local people often resent the transmigrants for the facilities and assistance they get from the government. The need for the settlers to develop a sense of community in order to help one another over the hard times conflicts with the goal of integration with the native population. The government now has a policy of allocating 10 percent of the places in each settlement to local people, but that will only draw the line between natives and settlers at a different point, not eliminate it. The official argument is that simply by increasing the population on the outer islands, transmigration improves the quality of life of the indigenous population. This is because bigger populations provide the critical mass needed to support markets, roads, schools, and economic diversification.

. . .

Lampung is Sumatra's southernmost province, a six-hour road-and-ferry trip from Jakarta, and the most popular destination for migrants to the outer islands. Indonesia's first official resettlement project in 1905 was in Lampung. By 1980, when the Suwendris arrived, 80 percent of the province's 4.6 million people were migrants, or descendants of migrants. The settlers have devastated Lampung. They have cut nearly all its forests, and in many areas repeated burnings have made the damage irreversible. The economic and social consequences of this overexploitation are now apparent. The first colonists settled on the relatively good soils of the flatlands. Their solid houses, surrounded by flowering shrubs and shade trees, line the main roads and speak of prosperity. These settlers have settled in; they have achieved the critical mass. Many own shops, work in offices, or work for the local government. They, not the small, native Malay population, are Lampung now.

Those who came later moved up the hillsides and onto the swamps. The steep slopes of Bukit Barisan, one of the most important water catchments for the rivers that irrigate the fertile plains of the province, are now almost treeless. Loggers took the trees, and the farmers who followed them used fire and knives to keep the regrowth down. Where farming proved impossible or unprofitable, families moved on, looking for a better place from which to make a living. Often they left behind only sterile *alang alang* grassland that no one knows how to make productive again.

A report by the FAO says that Lake Jepara, a major source of irrigation water, is filling with silt and cannot hold as much water as it used to. Telukbetung, the provincial capital, is suffering a growing number of water shortages as a result of siltation of riverbeds and accentuated dry seasons. Farmers cannot use many of the expensive irrigation systems that the government has built because there is not enough water in the dry season. When the rains come, the problem is reversed. The rivers, heavy with the silt of eroded mountains, run high, and the number of flash floods increases each year. In 1979 such a flood brought down a railway bridge spanning the Seputih River.

The government now considers Lampung to be dangerously overcrowded and is moving hundreds of thousands of new and long-time migrants to less populated areas. Many of them are not happy to have to move again, especially when they have doubts about the quality of the land to which they have been assigned, but officials are now seriously concerned about the effect of deforestation on water supplies. The government blames the problems in Lampung on the extent of uncontrolled immigration and on bad planning for government settlement schemes in the past. But new settlements elsewhere have similar problems.

Between 1980 and 1982, eighty thousand "official" transmigrants, mostly from Java, moved to a swamp forest between the rivers Suleh and Sugihan, east of the city of Palembang in southern Sumatra. There were serious problems from the beginning. "We did a study, but—partly because of the pressure of time—it was not so complete," Soedjino says. When I was there in October 1982, Air Sugihan, as the site is called, was suffering from poor soil, lack of water, a cholera epidemic, and a herd of marauding elephants.

There are no roads to Air Sugihan. The only approach is by boat from the coast or down the Suleh River from Palembang. The natural vegetation of the area is freshwater swamp forest. The riverbanks are lined with nibong palms, their huge, gracefully drooping fronds giving the scene the look of a painting by the Douanier Rousseau. These trees are in the same genus as the ones whose fossilized fruits have been found on the Isle of Sheppey, at the mouth of the Thames—

remains of the days when tropical vegetation covered southern England. Kites and hornbills fly overhead. Long-tailed macaques and silverleaf monkeys sit singly or in small groups in the branches of the trees that border the river. Deeper inside the forest, elephants forage for food.

I smelled Air Sugihan well before I saw it. For several miles before we reached the settlement, the air was heavy with the smell of smoke. The local transmigration chief, a twenty-five-year-old man called Trijoko Choiriyanto, who is responsible for one group of 2,600 families, told me that the worst of the fires was over. "People started fires in August to clear their land. It was so hot and dry then that many just spread out of control. In some cases the soil itself caught fire and burned the roots of the trees. Some of those trees are still standing, but they are dead." I assumed that the dense haze that hung over the newly cleared fields was a side effect of the eruption of Mount Galunggung in West Java in early 1982. The same cloud enveloped Palembang, Karangsari, and everywhere else I had been in Sumatra. Jakarta, too, and large parts of Malaysia, had been hazed over for several months. Later I learned that satellite pictures had revealed that the haze was caused by large-scale burning of trees in southern Sumatra and southern Kalimantan.

In theory—and this is what attracted government planners to these forests in the first place—such swampy soils should be ideal for growing rice, which is, after all, a swamp grass. First the forest is cut down, then canals are constructed. The canals have a dual purpose; they drain the swamps and also provide irrigation water, of varying degrees of salinity. Some swamps may have peat to a depth of fifteen feet, but as long as the peat layer is no deeper than five feet, such swamps may be suitable for growing rice. There are a number of transmigration settlements on such swamplands, with an average rice yield that is four or five times greater than yields in nonirrigated upland areas.

Because of the complicated hydrology of wetlands and its influence on soil quality, swamps are among the least well understood ecosystems for agriculture. The effects of different types of drainage on the soil and on rice yields are poorly understood. Local people have developed ways of using the resources of the swamps without damaging them. For centuries the Malay natives of Sumatra have made a living from the peaty soils and forests of the island. But their numbers have been low, and they have diluted their impact and spread their risks by engaging in a variety of activities, from fishing to rice growing to trading.

The courage and determination of the settlers at Air Sugihan is manifest. They have staked everything on making a success of this move. In the unrelenting, humid heat, they have tackled the back-breaking jobs of felling hundreds of huge trees, of digging up the dry, cracked soil and sowing seeds, of hauling water from the river to keep their precious crops alive. Scattered throughout the devastation of scorched stumps and rotting logs are small thriving patches of maize, taro, cassava, peanuts, sweet potatoes, and coleus. The multicolored leaves of the latter make it a popular houseplant in Europe and North America, though the settlers here grow it for food. In the rainy season they plant rice.

It is a tragedy that their toiling to replace the forest with annual crops may backfire after a few years. By burning fallen trees, the settlers get a better harvest in the short term because the nutrient-rich ash, which is basic, neutralizes the acid soil. But clearing and burning peaty soils impairs their ability to hold water and other nutrients. And water retention, rather than soil acidity, is the major problem of Air Sugihan's soils, according to Soeriaatmadja (who is known to one and all as Aat). Beyond the ranks of identical wooden houses scraggly remnants of forest, shrouded by the haze, mark a boundary between settlement units. Large expanses of land lay cleared and charred but as yet unplanted. The banks of drainage ditches are eroded into deep gullies by the rain. Some have caved in completely.

Compared with Karangsari, Air Sugihan seems more like a refugee camp than a village. There are no flowers, no trees planted to give shade or beauty, no shops, no market or gathering place—nothing to indicate a social or cultural life. Choiriyanto agreed that life in Air Sugihan was not only difficult but dreary. He hoped, he said, that some Balinese people would be assigned to this section and strengthen the social and artistic life of the largely Javanese Muslim community. But for now he has other, more immediate worries.

"Seventy percent of all the water at Air Sugihan is carrying cholera bacteria and eighty people have died so far," a transmigration official told me when I returned to Jakarta. "The study we did in the planning period wasn't so intensive. Part of Air Sugihan was settled under the plan-as-you-proceed system. So we need experts on climate, soil, water, and so on, to go down and make a comprehensive study. Then we will rehabilitate or whatever the term is." Droughts of the severity that had affected the water supply and caused the cholera epidemic occur every five or six years in that part of Sumatra. Nonetheless, he said, "We didn't expect it. We weren't ready for it."

Another thing that no one was ready for was the herd of elephants that lived in the forest of Air Sugihan. As the forest was cut down, the elephants, over one hundred of them, retreated into a small corner of forest that was left standing along the Suleh River. Striking out in search of food, the herd trampled crops, damaged buildings, and terrified the settlers. According to the local people, the elephants are particularly fond of bananas, coconuts, and rice. "People are in doubt about whether to plant out next season's crops," an Indonesian official told me, "because of the elephants." Aat showed me pictures of the devastation caused recently by one elephant in western Sumatra who had knocked down 170 homes. Elephants are a protected species in Indonesia. In Air Sugihan rumors flew that the settlers were to be shifted so that the elephants could be left in peace.

When he realized the scale of the problem, Salim formed a National Task Force on Elephants, and named Aat to head it. The task force decided to try to herd the elephants through the settlement to an as-yet-uncleared patch of forest thirty miles to the south. Their problem was that no one knew how to go about moving so many elephants. In a similar situation some years ago, Rwandan officials had to abandon a plan to move the last 146 elephants in the country twenty-five miles through a densely populated area because they could not find anyone in Africa who could run the drive. Instead the Rwandans immobilized the twenty-six babies of the herd and took them away by helicopter; the rest of the elephants the government "very reluctantly" shot. Sri Lankan elephant handlers told the Indonesians that a thirty-mile drive would take a year. The Sri Lankans move their herds in stages, allowing them to move freely back and forth between intermediate stages and gradually cutting off their retreat. But time was short at Air Sugihan. The task force couldn't allow the elephants to spend a year ambling through a crowded settlement.

The first meeting to discuss the settlement's elephant problem had been held in February 1982 at the Environment Ministry. In June the Indonesian cabinet gave its blessing to the drive, but it wanted the job done before the presidential elections in March 1983. That meant that the elephants would have to move in November, before the rainy season began.

The task force decided to cut a track that would funnel the herd through a corridor of forest running along the edge of the settlement to the elephants' new sanctuary 30 miles away. The route would take the herd over seven as yet unbridged canals. Two hundred soldiers, assisted by thousands of transmigrants, were to round the elephants up over a 150-square-mile area and drive them south using firecrackers, whirring chain saws, torches, megaphones, and tape recorders playing the sound of falling trees. The entire process was scheduled to take less than a month. Once the elephants were in their new sanctuary, called Lebong Hitam, the army would dig a trench, with a steep drop on one side and a solar-powered electric fence on the other, to keep them from returning to their old haunts.

The plan was far from foolproof. The forest corridor through which the elephants were to walk was so heavily logged over that it had disappeared in some places. In others, the way was littered with fallen logs, some four or five feet high, obstructions to both the elephants and their would-be drivers. The corridor was so close to the fields and houses of the transmigrants that makeshift barriers were needed to prevent the elephants from straying dangerously off-course. And if the drive succeeded, that would not be the end of the matter. Elephants elsewhere had overcome both types of barriers that the task force was planning to erect to prevent them from wandering back. One neutralized an electric fence by dropping an uprooted tree on it. Others had used their heads and tusks as bulldozers to turn a deep ditch into a gentle wrinkle in the landscape.

We were a small party at Air Sugihan a month before the drive was to begin: Aat; Graham Child, the director of Zimbabwe's National Park Service; Iain Douglas-Hamilton, who has spent many years observing elephants in East Africa; and I. My own study of elephants had been restricted to reading the Just So Stories, but then no one was looking to me for advice. Douglas-Hamilton and Child were there to advise Aat and other members of the task force on how to move the elephants. I had never seen Asian elephants in the wild, nor had Douglas-Hamilton or Child. We split up, with members of the task force, to look for the elephants, each small group of searchers gathering a large following from the children of the transmigrants along the way. At the end of the day Child and I were cursing our luck and Douglas-Hamilton was waxing lyrical about the placidity of the beasts, who had allowed him to come within six yards of them. "It was magnificent. They charged us a few times, but nothing serious. Otherwise, they let us stay near them for a couple of hours. Apart from the mountain gorillas in Rwanda and Zaire, I don't know anyplace where you can get as close as that to a large wild mammal. You shouldn't have turned off. We found them about fifteen minutes after you left us," he said, in a tone that was probably not meant to sound like rubbing it in.

"I've never heard of anyone driving as many as one hundred elephants at one time," Douglas-Hamilton told the task force when we returned to Palembang. Still he and Child were able to give some advice based on their experiences with African elephants—and a few words of warning. They suggested pushing the elephants ahead by manipulating the down draft from helicopter blades. Child also advised putting water in the barrier moat, as is done in Africa, so that elephants butting the electric fence get a stronger shock. Child told a moral tale about the difficulty of monitoring animals in the wild. The manager of a certain roe deer reserve in Zimbabwe allowed hunting under very strict conditions. His three game managers always kept an exact count of the roe deer population. At one stage when there were 89 deer in the reserve, they decided to kill them all in order to replace them with better stock. At the end of the cull there were 249 dead deer.

Both Child and Douglas-Hamilton were worried that in rushing to beat the rains, the task force was not allowing itself enough time for preparation. The Indonesians wanted to start within a month, but Douglas-Hamilton and Child urged them to consider delaying it so that the army could do a pilot run and the local people could be better informed about what to expect. Aat objected. "No, there is a momentum that would be lost with postponement. The commander-in-chief of the army is very ready to help now. And the people here are already so worried about the elephants that it would be a mistake to delay." Another task force member, a young Indonesian zoologist named Jack West, added that if they waited until next year and the logging went on as it had, there would be no trees left to keep the elephants on their trail.

The army moved into Air Sugihan on November 15. Two hundred soldiers and an equal number of civilians pushed a rough track through to Lebong Hitam. Rounding the elephants up and trying to count them was a hellish job, much more difficult than anyone had foreseen because of the vast area, the confusing vegetation and terrain, and the impossibility of labeling the beasts or keeping track of them once they had been found and counted. The roundup took two weeks, and the army found it had 232 elephants to herd. Several thousand transmigrants helped drive the elephants, but progress was slow. The animals, confused and frightened, let out wrenching bellows and sometimes charged their drivers. The soldiers had orders not to shoot except as a last resort to save human life. It took forty-four days, rather than the scheduled twenty, to travel the thirty miles to Lebong Hitam. "It was, of course, forty-four hardworking days, not only for the army but also for the transmigrants who got involved," Aat said later. "And it was just like a miracle because nobody, none whatsoever, got hurt during the whole operation. Right now the elephants are safe in their new habitat." The place had been designated as a nature reserve, and Aat was charged with setting up a management plan for it.

Before it was turned into a home for elephants, Lebong Hitam was being prepared for the next wave of transmigrants to Air Sugihan. Part of it had already been cleared for the construction of a large drainage and irrigation canal. Immediately to the south of Lebong Hitam, the land has been allocated to more transmigrants, though clearing has not yet begun. The twenty-five-thousand-acre reserve will be surrounded by transmigration settlements. "Everybody says it's too small," Jack West said. "But we hope that once the elephants settle in, it can be extended to 100,000 acres."

In the spring of 1983 I wrote a piece about the elephant drive for *New Scientist,* a British magazine. Soon after, I had a letter from Tony

Whitten, an English ecologist working for the University of Sumatra's Center for Environmental Studies. It enclosed a clipping from *Kompas,* Indonesia's major daily newspaper, dated April 4, 1983. Headlined "Returned Again," it read:

> Operation Ganesha was successful at the end of last year in driving 232 elephants from the Air Sugihan Transmigration Area to Padang Sugihan, both in South Sumatra. From February, however, a large group of these elephants have returned to the complex called Canal 20, Air Sugihan, and disturbed the transmigration settlement and the fields. The photo of the elephants, above, was taken in March and represents a part of more than 100 wild elephants which have returned to their original home. Clearly it is not possible to stop these large beasts from crossing between the protected forest and Padang Sugihan just by using ditches. Maybe the ditches weren't built properly. Man-made barriers aren't really capable of preventing the return of the elephants.

. . .

Karangsari, though certainly thriving compared with Air Sugihan, is not without its problems. The soil along the Penet River, which forms the southern boundary of the reserve, where most of the farming villages are, appears capable of supporting subsistence agriculture, but most of the reserve has marginal soils that are unfit for agriculture. If people penetrated deep into the reserve and tried to farm there, they would probably destroy the soil beyond hope of reclamation. The few remaining trees in the reserve are under pressure from villagers collecting wood for fuel. A herd of elephants lives in the reserve. Forty years ago they could walk across Sumatra, from the mountains in the west to the coastal swamps in the east. Now they are isolated in a sea of human settlements and grasslands. The villagers are afraid of the elephants and worried about the damage they do to crops. Periodically the people of Karangsari band together, brandishing torches and ringing bells, to drive the elephants away from the village. Villagers also kill elephants, tigers, and other rare animals that they regard as valuable or dangerous. Fishermen have overcropped the nibong forest along the coast, cutting the sturdy stems to build offshore fishing platforms. Fishing along the coast interferes with the swampy feeding grounds upon which, especially in the dry season, elephants, sambar deer, tapir, and other animals depend. But, for the villagers, the biggest problem is that the government has ordered them all to leave.

For years the Department of Forestry, and the Directorate of Nature Conservation (PPA), itself a division of the Forestry Department, worked for contradictory goals—one to exploit, the other to protect, the forest. The timbermen won most of the early rounds, but in 1979 the PPA prevailed upon the national government to ban logging in the reserve. In 1971, two years after the governor of Lampung Province granted Karangsari legal status as a village, the minister of agriculture overturned that decision on the advice of the PPA, which also asked that the villagers be relocated. In 1974 the Forestry Department canceled the timber licenses, though logging continued illegally for another five years, during which time several new fishing and farming settlements sprung up on the edges of the reserve. Now that the reserve is virtually deforested, conservation has been given top priority.

In 1979 the PPA commissioned a study of Way Kambas from the United Nations. The authors concluded that "most of it has been destroyed during the last 20 years, especially in the period of 1968–74, when the Reserve was opened to commercial timber exploitation." They reported that "the reserve has been irreversibly changed and is no longer a representative example of a natural ecosystem. . . . It is not likely that within a relative short period of a few centuries the original forest type will be able to come back in the major part of the reserve." Nonetheless, they felt that "it could still have high conservation value as a reserve for species such as elephant, tiger, and other important animals." But Way Kambas should be a game reserve, not a national park, the authors argued. "The concept of national park would be completely devalued if such a disturbed area becomes so classified." To give the reserve a chance to recover, the authors wanted several hundred families from the fishing villages resettled and the boundaries of the reserve redrawn to exclude the remaining, mostly agricultural, settlers. The government decided to go beyond those recommendations. All five thousand families living within the existing border of the reserve are to be moved to North Lampung by 1984.

Wayan Gatar, Karangsari's schoolteacher, approves of safeguarding endangered species. "It is very good to protect animals—for study, research, culture. But I don't like to leave Karangsari. It is safe here, we are with our friends, and we get good crops." I asked Gatar if he had hopes that North Lampung might turn out to be satisfactory. "Not yet," he said softly.

Another man, who came to Karangsari in 1963, expressed it more strongly. "We have been here for nineteen years and we don't want to leave. The soil in the new place, in North Lampung, isn't as good as this and we are afraid of being poor there. Also the people here are friends and we don't know if we can stay together when they move us." I asked one man if he had protested to the officials who told him the village was to be disbanded. Everyone in the room laughed at the question. "No," he said. "Not brave. I was afraid. Democracy in Indonesia is not the same as democracy in the U.S.A."

Ketutgede Suwendri and his wife have already been relocated; the day when they will be finally settled, when they can send for their children to join them, has been set back once more. Before he left, he told me, mildly, that it was "a little unfortunate" that they should have to uproot themselves again, to abandon their newly planted fields. He would try hard, he said, wherever the government put him. He had heard the government would help them. Perhaps the new land would be good.

12
Fever Bark

In 1735 a thirty-one-year-old French botanist, Joseph de Jussieu, set out for Latin America as part of the first non-Spanish expedition to be allowed to enter the continent. The main object of the small group was to determine the circumference of the earth by taking measurements at the equator, but de Jussieu had his own goal—to find and describe the cure for a disease that had killed more people than any other in history. His account would, he hoped, raise him into the charmed circle of the Académie Française.

For centuries, malaria had struck many hundreds of millions of people each year and killed several million of them. Alexander the Great was one of its well-known victims, and its debilitating effects have been linked to the demise of ancient Greece and the Roman Empire. The chief treatment in Europe was bloodletting, which added its mite to the death toll. In 1633 Father Calancha, a Jesuit priest who lived in Peru, learned of a cure for malaria. He wrote: "A tree grows which they call the fever tree in the country of Loxa [now Loja in Ecuador] whose bark is the color of cinnamon. When made into a powder amounting to the weight of two small silver coins and given as a beverage, it cures the fevers. It has produced miraculous cures in Lima." His news was ignored in malaria-ridden Madrid, Lisbon, Rome, London, and Paris. Twelve years passed—and another twenty or thirty million people died of malaria—before the bark reached Europe.

It was another Jesuit, Father Bartolomé Tafur, who finally brought the fever bark—then also known as Peruvian bark or Jesuit's powder—to a conclave in Rome in 1644. The business of choosing a pope in malaria-plagued Rome always claimed the lives of several high churchmen. But Tafur's cure worked, and soon Jesuit missionaries in South America were sending the bark, various species of the genus *Cinchona,* back to Rome and Spain regularly. The 1655 conclave, during which Alexander VII was named pope, was the first in the history of the Church with no deaths from malaria.

Jesuit's powder met with violent opposition from the medical establishment and the public. So feared were papists in general and the Jesuits in particular throughout Protestant Europe that there were demonstrations in London against the bark. Jesuit's powder, the crowd believed, was part of a plot to exterminate all Protestants. In 1658 Cromwell died of malaria rather than accept the cure of his papist enemies. Doctors, whatever their religious views, treated the cure with the scorn that the profession continues to bestow upon folk remedies.

But the scourge of malaria created sufferers who were desperate enough to try anything. One of these was King Charles II, who in 1678 was cured of malaria by Robert Talbor, a fashionable London healer with no medical training. Talbor refused to reveal the secret of his malaria cure. His strong denunciation of Jesuit's powder was not enough to endear him to the medical profession, which reacted with horror when Charles knighted him and forced his admission to the Royal College of Physicians. Talbor went on to become a darling of the French court after curing another royal patient, the son and heir of Louis XIV. The French king gave Talbor a pension for life, and paid him three thousand gold crowns for the secret of his cure, promising to keep it under lock and key until Talbor's death. Talbor's popularity grew with his repeated success in curing malaria; his secret remedy was welcomed by all but the doctors.

When Talbor died suddenly in 1681, Louis revealed the secret remedy. It consisted of six drachms of rose leaves, two ounces of lemon juice, and a strong infusion of Peruvian bark, administered in wine. There could no longer be any doubt that Jesuit's powder was the cure for malaria. Suddenly doctors were vying with one another to announce their longtime faith in the cure. One convert who had earlier denounced the bark, Thomas Morton, said, "Opposition to Peruvian bark was mainly a result of a conspiracy between physicians and apothecaries who resented the cure of a disease which had been for so long an unmixed financial blessing."

Cinchona was suddenly in great demand. It revolutionized the art of medicine, said the seventeenth-century Italian physician Bernardin Ramazzini, as profoundly as gunpowder had the art of war. But since no one knew how or why it cured malaria, or which species yielded the best bark, the quality varied from shipment to shipment. And as the demand for cinchona grew, so did the tendency to mix real cinchona bark with other ineffective or even dangerous barks. It was hard to guard against this, since, apart from a few Jesuits who organized Indians to collect the bark, no European had seen the fever bark tree.

Jussieu was determined to change that. On arriving in Ecuador, he parted from his companions, the astronomer Pierre Bouguer and the mathematician Charles Marie de La Condamine, and headed south alone. (Bouguer and La Condamine spent the next seven years in the mountains near Quito trying to measure the equator, and quarreling furiously. So soured did their relations become that Bouguer one day abruptly returned to France, leaving his colleague working on another mountain peak for two and a half months before coming down to discover that there was nothing to check his measurements against.) Aiming for Loja, Jussieu walked three hundred miles along the Andes in search of the fever bark tree. The cinchona tree is attractive. Richard Spruce admired it as "a very handsome tree . . . in looking out over the forest, I could never see any other tree at all comparable to it for beauty." But cinchonas are not easy to find. Instead of growing in convenient clusters in the lowlands, they are scattered across mountainsides in the wettest parts of the Andean rainforest—in Ecuador, Bolivia, Colombia, and Peru. (William Steers, a U.S. government botanist who collected cinchona bark in Colombia during the Second World War, described the climate: "In the dry season it rained only in the afternoon.")

Jussieu spent almost thirty years in the forest, living among the Indians and studying, collecting, and writing detailed reports on the different species of fever bark tree he found, and on how best to cultivate the trees and collect and preserve the bark. His hopes of sending back the first eyewitness account of the tree were dashed when La Condamine passed briefly through Loja in 1737. The mathematician got a glimpse of the tree and posted some quick notes off to France. The next year, when his description was read at the Académie, La Condamine was hailed for his contribution to knowledge.

But Jussieu's exhaustive studies and unrivaled collection assured him of a triumphant reception when he returned to Paris. In 1761 he was in Buenos Aires, preparing to go home. The night before the ship was to sail, Jussieu left all his material in the care of a servant with strict instructions to guard it well because of its great value. The next morning both servant and baggage were gone, never to be seen again. Jussieu collapsed under this calamity. He wandered in Latin America for ten more years before returning in 1771 to Paris, where he was committed to an asylum and died, never having recovered from the shattering of his dream.

A century and a half had passed since the first news of cinchona had reached Europe, and apart from La Condamine's brief layman's description, there was still no accurate scientific information about the tree. Many people tried to find the tree, to study it, and to take it back to the Old World, but their efforts were dogged with disaster. La Condamine himself had some seedlings collected to be sent back to Europe; incredibly they survived the hazardous three-thousand-mile journey down the Amazon only to be washed overboard as they were being transferred to a seagoing vessel.

In 1761, the year of Jussieu's disaster, José Celestinto Mutis, a young Spanish doctor, took up the position of physician to the viceroy in what was then the New Kingdom of Granada, now the Republic of Colombia. Mutis, like Jussieu before him, was burning with the ambition to become the world's authority on cinchona. He did some work on the tree, but was frustrated by his court obligations. In 1782, at the age of fifty, Mutis finally abandoned the court for the forest. He returned only when the king of Spain agreed to make him the head of a new Botanical Institute with headquarters in Bogotá. Mutis had a staff of scientists working under him and the institute was well equipped. Alexander von Humboldt, the German naturalist-explorer, visited it and said that its library was the best in the world for natural sciences. But in ten years Mutis produced nothing. After his long years in the tropics, trying to do two jobs at once, he was showing signs of a nervous breakdown; he was increasingly erratic, quarrelsome, and neurotically protective of his specimens. Under pressure from inspectors sent by the king, Mutis produced two short papers in 1793. Although they contained nothing new, they were the first publications on cinchona since La Condamine's notes half a century earlier. Mutis claimed to have detailed information on the natural history of cinchona and on how to choose and prepare the bark, but he refused to reveal it. His papers and specimens were deliberately arranged to hide what he had learned.

Just before his death in 1808, Mutis told his successor, a talented scientist named Francisco José de Caldas, how to make sense of what he was leaving behind. Caldas set about sorting out Mutis's papers, but gave it up midway to join Simón Bolivar's fight for independence from Spain, in which cause he was captured and condemned to death. He accepted the sentence but asked for six months to finish the work to which only he held the key, offering to work in chains. His request was denied and he was shot in the back by a firing squad on October 29, 1816. The material he and Mutis had collected over more than fifty years was shipped back to Spain to gather dust in a shed at the Madrid Botanical Gardens. Almost two hundred years after the Jesuits had learned of its powers, there was still no proper study of cinchona.

While Mutis was still at court, another Spanish expedition had gone out in search of cinchona. Three men—Hipolito Ruiz, José Pavón, and José Dombey—labored for eleven years collecting different types of bark to be analyzed by scientists in Madrid. The entire cargo was lost in 1788 en route to Spain, when the ship was attacked by English pirates. It turned up in 1852 in the British Museum, whence it had come via an auction house.

It was von Humboldt, together with the French botanist Aimé Bonpland, who gained the distinction of publishing—in the early part of the nineteenth century—the first serious study of the different species of cinchona. Soon after, in 1820, two French doctors, Joseph Pelletier and Joseph Caventou, discovered the alkaloid that made cinchona bark effective against malaria. They called it quinine, after the Indian name *quinaquina,* meaning bark of barks.

By then more and more voices were being raised to warn of the dangers of wiping out the trees by overcollecting. Collectors either cut down the trees or stripped off so much bark that they died anyway. The Jesuit-inspired practice of planting five trees for every one stripped had died away. In 1735 one prescient Spaniard had written, "Though the trees are numerous they have an end." And Humboldt warned that "if the governments in South America do not attend to the conservation of the quina . . . this highly esteemed product of the New World will be swept from the country." The only conservation measure was a strict ban on exporting seeds or seedlings, designed to preserve the South American monopoly.

With millions dying of malaria in India and the Dutch East Indies every year, both Britain and the Netherlands chafed against the South American monopoly. They each wanted their own secure supply of quinine. In the mid-nineteenth century both countries financed expeditions to the Andes to collect material to be grown in their own tropical colonies. Several Dutch ventures resulted in two million trees being grown in Java, none of which contained a useful amount of quinine.

For the British government, Richard Spruce and other botanists collected many hundreds of thousands of seeds and seedlings of different species of cinchona. They had to be collected secretly and smuggled out, since all the countries in which cinchona grew naturally had banned the export of seeds or living plants. Spruce's collection reached India and Ceylon safely, was planted, prospered, and produced so little quinine as to be useless.

As the British and Dutch had learned to their cost, the concentration of quinine in the bark varies widely between different species of cinchona, and between different populations of the same species. That the European powers were able to develop a secure supply of good-quality cinchona, thus removing one of the main obstacles to Europe's domination of the tropics, was ultimately due to a Bolivian Indian who put friendship above patriotism. Manuel Incra Mamani was a bark collector, who for eighteen years worked in Peru for an English bark-and-fur trader, Charles Ledger. In his years as a middleman, Ledger had noticed that the strongest bark seemed to come from trees with scarlet leaves, called *roja* by the Indians in the region. The exact locations of the best stands were closely guarded Indian secrets.

In 1861 Ledger asked Mamani to lead him to the *roja.* Mamani, who knew of a particularly powerful stand near the Rio Beni in Bolivia, refused. He left Ledger's employ but returned four years later, bearing the seeds his friend wanted. Ledger paid Mamani $750, but in 1874, after Mamani had returned to Bolivia to collect more seeds, his employer wrote, "Poor Manuel is dead also; he was put in prison by the Corregidor of Coroico, beaten so as to make him confess who the seed found on him was for; after being confined in prison for some twenty days, beaten and half starved, he was set at liberty, robbed of his donkeys, blankets, and everything he had, dying very soon after."

No one knew at the time that *C. ledgeriana,* as the new species was eventually called, would turn out to be the most potent type of cinchona ever found, with up to 13 percent quinine in the bark, compared with an average of 2 percent in the trees collected on official British and Dutch expeditions. Those seeds, the source of quinine used throughout the world, saved millions of lives. Ledger was a visionary, not simply out for profit. He wanted to save lives and bring glory to the British Empire. He offered the seeds to the British government, through his brother George in London, but—to his great disappointment—was turned down. The British did not yet realize the worthlessness of their expensive cinchona-collecting expeditions. Fearing that the seeds would die, George Ledger turned to the Dutch government and was able to persuade one minister to buy one pound for about $40. Then in desperation Ledger actually resorted to peddling the rest of the seeds in the streets of London, finally managing to unload them on a British planter on home leave. But the planter later lost his nerve and traded them for a more "reliable" variety at the British cinchona plantation in India.

In this way the British finally came into possession of the most valuable seeds in history. But the managers of the Indian plantation, who had already successfully germinated several million worthless cinchona trees, failed to grow to maturity any of the more than one million "Ledger" seeds they had fortuitously obtained. The Dutch were luckier with their one pound; within a few years they had twelve thousand "Ledger" trees growing in their Javanese plantations, and possessed the basis of a century-long monopoly on the most important medicine in the world. In 1917, when Sir William

Osler outlined the remedy for malaria, he wrote: "Treatment. This is comprised in one word: quinine."

. . .

The Dutch monopoly was interrupted in 1942, when the Japanese occupied Java. In 1940, at the beginning of the Second World War, the Germans had taken Amsterdam—and all the quinine reserves in Europe. U.S. Army memos written in 1942 show the extent of malaria's threat to the war effort in the Philippines: "20 percent of fighting units are not effective due to malaria and 50 percent of some units have subclinical malaria and are considered potentially ineffective." Luckily for the Allies, an American forester named Arthur Fischer had in 1917 convinced the U.S. military governor of the Philippines to pay $4,000 to an English captain for smuggling two Horlick's Malted Milk jars full of cinchona seeds from the Dutch plantations on Java to the Philippines. Fischer started a small cinchona plantation with the seeds. In 1942 Fischer, then a colonel in the Philippines and himself stricken with malaria, and weighing 96 pounds instead of his usual 150, led a daring sortie to his old plantation, sending seeds and bark out on the last emergency rescue flight, just after the fall of Bataan.

Those seeds did not mature in time to provide the Allies with wartime quinine. Instead, the U.S. government's Board of Economic Warfare sent a team of botanists led by William Steere, later head of the New York Botanical Garden, and Raymond Fosberg, now at the Smithsonian Institution, to Colombia. The 12.5 million pounds of dried bark they collected was the Allies' mainstay throughout the war. New plantations were also established in Africa, Peru, and Mexico. In 1944 two American scientists, William Doering and Robert Woodward, synthesized quinine, just too late to make up for the Allies' desperate shortage of quinine caused by Japan's occupation of Java. But the era of synthetic malaria cures had begun. Medicines such as chloroquine and primaquine, based on the structure of natural quinine, were cheaper to make—and sometimes more potent—than quinine.

Today, however, there are serious problems with the synthetics. New strains of malaria parasites have developed that are resistant to all known quinine substitutes. In many areas this means that the disease, which was once thought to be on the wane, is increasing. In 1952 there were one hundred million cases of malaria in India; quinine and its derivatives reduced that to sixty thousand in 1962. Fourteen years later six million Indians had malaria. In Thailand the two main anti-malarials have been chloroquine and fansidar. In 1960 more than 90 percent of people treated with choloroquine were cured; in 1982 fewer than 20 percent so treated recovered. Resistance to fansidar is also growing. In 1976 malaria killed 1.5 million people worldwide, according to the World Health Organization.

Doctors are turning back to quinine, usually in combination with the antibiotic tetracycline. Synthetic drugs work against one or a few strains of malaria, while extracts of cinchona bark—which contains at least thirty alkaloids besides quinine—have a broad-spectrum anti-malarial action. Malarial parasites have not developed a resistance to extract of cinchona, though it has been used widely for centuries. With natural cinchona again in demand, abandoned plantations in Zaire, Guatemala, and Java were brought back into production in the 1960s.

Scientists all over the world are now searching for new antimalarial drugs. But they are trapped on a treadmill of continually evolving disease strains requiring a continuous supply of new drugs. In this battle cinchona is important as a medicine in its own right and as a model for new synthetics. Other plants in the forest may serve the same functions. In Colombia, the Institute of National Health is testing four plants that doctors hope may be new natural cures for malaria. "We have to face the problem that in the future, as resistance develops, there may be no effective anti-malarial drugs," Carlos Espinal, the head of the institute's Department of Immunology of Malaria, told me. "A group from the biology department lived with the Choco Indians and learned which plants they used for acute fever. We've narrowed the field down to the four most promising species. The Indians found the cure for malaria once; we hope they will help us a second time. The Indians know more than we do about such things."

13

Fruits of the Earth

As we destroy the forest, millions of species of plants and animals, the vast majority of which are completely unknown to science, lose their habitats. Scientists have scarcely begun to ask how man might benefit from the products of the forest. Fewer than 1 percent of tropical forest species have been examined for their possible use to mankind—that is, screened for even one type of chemical compound. Yet one can judge their potential by the effect rainforest species have already had on medicine, agriculture, and industry. It is no exaggeration to say that without quinine, coffee, and rubber their histories—and mankind's—would be different.

Rainforest plants tend to be rich in what are called secondary compounds, chemicals that help them to survive but are not directly essential to growth. Some are poisonous, others produce warning colors, still others attract pollinators or "guardian" species that repel predators. Insect predators respond by evolving resistance to the toxic compounds their host plants produce. Many plants and animals survive by exploiting difficult ecological niches. They have evolved characteristics that, for instance, enable them to withstand toxic chemicals, grow quickly, or tolerate strong heat and humidity. To us all these elaborate defenses and adaptations simply make some species taste good, others highly nutritious, others poisonous, and still others useful sources of industrial oils, dyes, resins, and so forth.

Seventy percent of the three thousand plants identified by the U.S. National Cancer Institute as having anti-cancer properties are rainforest species. Studies have shown that the tropics are twice as likely as temperate regions to produce plants that contain alkaloids, an important group of physiologically active chemicals. Plants, or their active ingredients, can be used directly as medicine, as "building blocks" from which chemists develop semi-synthetic compounds, as blueprints for the synthesis of natural compounds, or perhaps to create tumors or poisons on which other compounds are tested.

Much modern surgery depends upon the bark of a South American liana, discovered by the Rayas Indians, who use it as an arrow poison. In 1541 a group of Portuguese, led by Francisco de Orellana, became, by accident and without realizing it, the first Europeans to travel the whole length of the Amazon River. They were also the first to record the deadly effects of this poison, curare. When a member of the group was hit in the hand by an Indian arrow, Orellana wrote: "The arrow did not penetrate half a finger, but, as it had poison on it, he gave up his soul to the Lord."

For centuries explorers and naturalists tried to find out what was in curare, a task made more difficult by the fact that different tribes had different recipes, some calling for as many as thirty ingredients. There are three main sorts of curare, each based on a different liana. In 1800, von Humboldt, visiting Esmeralda on the Orinoco River, gave the first detailed description of curare being prepared.

> We were fortunate enough to find an old Indian less drunk than the others who was occupied with the preparation of curare poison from the freshly collected plants. This was the local chemist. We found at his dwelling a great clay cauldron for soaking the vegetable juices, some shallow vessels for evaporation, some plaintain leaves rolled into cones, serving to filter the liquid. In the cabin, which was transformed into a chemical laboratory, the greatest order prevailed. The Indian was known in the mission by the name of "maître du poison." He had the impassive air and pedantic tone formerly found in the European pharmacist. "I know," said he, "that the whites have the secret for making soap and this black powder which, if one misses, has the

> fault of making a loud noise that scares away the animal. Curare which we prepare is far superior to that which you make over there, beyond the seas. It is the juice of a plant which kills quietly without one knowing whence the blow came."

The first proof that curare could be a boon to medicine came from a donkey who responded to artificial respiration. Charles Waterton, the eccentric Victorian explorer (a passionate Catholic, he made his home into a museum with such exhibits as "The English Reformation Zoologically Demonstrated," in which preserved reptiles were contorted into portraits of famous Protestants), was also one of the few Europeans who have ever seen curare being made. He brought some back to London from what is now Guyana and in 1814 injected a donkey with it. Within ten minutes her heart stopped. Waterton cut a hole in her windpipe and inserted a pair of bellows as a way of giving her artificial respiration. Soon she "held her head up and looked around," but Waterton had to keep pumping for two hard hours before she was fully recovered. The experiment showed that curare killed by immobilizing the muscles so that the victim stopped breathing. As long as they made sure the patient kept breathing, doctors could use curare to relax tense muscles.

Today d-turbocurarine and other alkaloids isolated from the various curare lianas (different species of the genera *Chondodendron* and *Strychnos*) are used to treat multiple sclerosis, Parkinson's disease, and other muscle disorders. Without them such delicate operations as tonsillectomies and eye and abdominal surgery would be impossible. The antidote to curare is neostigmine, a chemical derived from the Calabar bean, a West African plant that is also used to treat glaucoma and as a blueprint for certain synthetic insecticides.

Because researchers at the Eli Lilly company in Indianapolis and at the University of Western Ontario took the claims of folk medicine seriously enough to investigate the Madagascar periwinkle, there is now a 99 percent chance of remission in case of lymphocytic leukemia—a disease that mostly affects children. Fifty-eight percent of Hodgkin's disease sufferers survive for ten years after treatment, compared with 2 percent in 1961. Scientists have found sixty alkaloids in this one plant: Some reduce blood pressure; others lower the concentration of glucose in blood. But the two most important are the anti-tumor agents vincristine and vinblastine, which have revolutionized the treatment of leukemia and Hodgkin's disease. It takes fifteen tons of periwinkle leaves to make one ounce of vincristine, which sells for $100,000 a pound. Vincristine alone has annual sales of over $50 million.

Scientists at the U.S. National Cancer Institute had already tested the Madagascar periwinkle, but their screening program failed to detect the plant's potential. It is impossible to devise a screening program that will pick up everything about a plant. Traditional knowledge can be very helpful in narrowing down the field of inquiry to those species most likely to be useful. "If this plant had not been analysed," John Humphreys, a British biochemist wrote in 1982, "not even a chemist's wildest ravings would have hinted that such structures would be pharmacologically active." But the National Cancer Institute hopes to find more such plants. "Simply based on statistical probability, I reckon there are nine or ten superstars out there, just as important as vincristine and vinblastine," says James Duke, who is head of the U.S. Department of Agriculture's Germplasm Resource Laboratory, which provides the institute's scientists with the plants they analyze.

Some rainforest plants, such as the South American plant ipecac, have been used medicinally for years. Ipecac, which cured Louis XIV of amoebic dysentery, is still the most effective treatment for that disease. But it appears that there are many medicinal plants still to be discovered. A survey of 1,500 Costa Rican plants indicated that 15 percent had potential as anti-cancer agents. In 1979 the Madagascar periwinkle was the only major anti-cancer drug from tropical rainforests. Now dozens are reaching their clinical trials.

Steroid hormone medicines, such as cortisone and diosgenin, the active ingredient in birth control pills, were developed from wild yams (several species of *Dioscorea*) in Mexico and Guatemala. Biochemist Norman Applezweig, a leading authority on steroid chemistry, wrote in 1977 that without "Mexican diosgenin as the most versatile and available steroid raw material it is probable that we would not have had oral contraception in our time, or even in our century." The shrub *Rauwolfia serpentina* has been an important tranquilizer in India and Southeast Asia for thousands of years. Twenty years ago Western scientists recognized its value, and its derivative, reserpine, is now widely used to treat hypertension. The heart medicine strophanthin, an important alternative to digitalis because it has fewer gastrointestinal side effects, comes from a wild West African vine, *Strophanthus gratus,* long used locally as an arrow poison. Among forest plants recently found to have medical potential are the greenheart tree, which has a reputation in the Amazon as an effective female contraceptive. Another is *Clibadium sylvestre,* a plant from which Guyanese tribes have derived a powerful fish poison known as cunaniol that may be a great aid in heart surgery.

Tetrodotoxin, derived from several species of Central American frog, is an anesthetic 160,000 times as strong as cocaine. Haitian voodoo doctors have long taken advantage of its power to lower the metabolic rate drastically enough to give the appearance of death. Tetrodotoxin is the main ingredient in the poison they use to create zombies. After the victim is buried, the voodoo priest "revives" him (performs certain magic ceremonies meant to coincide with the natural wearing off of the effects of the drug) and feeds him a paste based on another plant, *Datura stramonium,* a close relative of jimson weed. Datura is a powerful hallucinogen that keeps zombies, who may be sold as slaves to work on sugar plantations, in a permanent state of catalepsy.

Seven thousand medical compounds in the modern Western pharmacopeia are derived from plants. One quarter of all prescription drugs in the U.S. contain one or more of those compounds. Norman Farnsworth of the University of Illinois estimates that the value of plant-derived prescription drugs in the United States alone was $8 billion in 1980. The figure for nonprescription drugs is much higher; Americans bought $165 million worth of plant-derived laxatives alone in 1980. If drugs derived from animals and microbes are included, the U.S. spent $20 billion on medicines with a natural component in 1980. For the industrialized world as a whole, the value of such medicines, according to Farnsworth, was at least $40 billion.

. . .

A number of endangered primates of the rainforest are used in medical research. The African green monkey is the only source of virus-free polio vaccine; the owl monkey of Central and South America is the only known animal suitable for research into human malaria; chimpanzees, of which there are only fifty thousand left in the wild, are the only animals on which anti–hepatitis B vaccines can be tested. But, although captive-bred animals have the advantages of a known pedigree and disease history, they cost more than wild animals. So most primates used for medical research are taken from the wild—fewer than 10 percent of those imported into the U.S. are from captive breeding centers.

. . .

In developing countries, where modern drugs are in short supply and local traditions still strong, natural medicines are the only sort most people will ever use. Many peasants who have come to the Amazon from the crowded cities of Northeast Brazil have switched over to locally available medicinal plants, calling them by the brand names of the aspirins and cold remedies that they used to buy. Many if not most tribal people can name thousands of plants. They can distinguish between groups identical to taxonomists and can control the strength of the medicines they make. People in India use close to 1,000 wild species as medicine. The traditional Thai pharmacy uses 500 medicinal plants. At least 150 of these have never been scientifically identified or analyzed. In the Malay Peninsula 2,432 species of plants, one third of all the species there, are used by the local people. In a lifetime of collecting in the Amazon, Richard Evans Schultes, professor at Harvard University and director of its Botanical Museum, has found more than 1,000 plants that Amazon Indian groups use medicinally, including 6 that he believes may be effective contraceptives. Worldwide, tribal people use at least 3,000 plants to control fertility. In 1977 the World Health Organization said that investigating these would be the best way to find new methods of birth control.

If even a small proportion of these plants turned out to be as useful as the wild yam, Madagascar periwinkle, or cinchona, any investment in finding—and protecting—them would be justified. The search is difficult because scientists do not know what they are looking for—any plant could in theory contain compounds that might have any effect on any condition. It may help to concentrate on plants that local people have identified as being particularly powerful. But gaining the confidence of local healers requires time and good faith. "Some companies," says Farnsworth, "just can't take the aggravation."

In any case, many scientists and pharmaceutical company executives prefer a high-technology image. The chemist R. S. De Ropp notes "the rather contemptuous attitude which certain chemists and pharmacologists in the West have developed toward both folk remedies and drugs of plant origin." This, he says, is to make "the error of supposing that because they had learned the trick of synthesising certain substances, they were better chemists than Mother Nature who, besides creating compounds too numerous to mention, also synthesised the aforesaid chemists and pharmacologists."

"A lot of pharmaceutical companies are not even willing to let people know that their medicines come from plants," Edward Ayensu, who is the head of the Smithsonian Institution's Office of Conservation, commented to me. Ayensu is a striking figure, impeccably tailored, whose jet black skin is set off by a dazzling toothful smile, a wafer-thin gold watch, and a plain gold ring. I first met him at a meeting in London to mark the tenth anniversary of the 1972 United Nations Conference on the Human Environment—the one that first focused popular attention on the destruction of the rainforests. Ayensu, Ghanaian by birth, commutes between his Smithsonian office in Washington, D.C., and Paris, where he heads the International Union of Biological Sciences. But those are his home bases, so he doesn't really consider transatlantic hops as traveling. Delhi, Moscow, Nairobi, Cairol, Manila, and Peking, however, do count. Against stiff competition, Ayensu is a front-runner for the title of eco-jet-setter of any particular year. I myself entered the junior sweepstakes in the course of writing this book, and rarely have I been in a place where Eddie Ayensu was not, or had not just been, or was not just about to appear. Luckily for him, he has developed a technique, involving starvation and sobriety, for withstanding the rigors of long-distance flights that enables him to preach the conservation gospel wherever he is.

The laxative Serutan is, as its advertisements told us for years, "nature's" spelled backward. It is also the husk of the seed of a common plantain, *Plantago ovata.* The husk has a semi-permeable membrane that helps cure constipation by attracting water. "In the United States," says Ayensu, "twelve ounces of Serutan costs $4.98. In India the raw material from which it is made sells for less than half a cent per pound.

"What is modern medicine, anyway?" Ayensu asks. "Western medicine is as crude as herbal medicine. Do you know what aspirin does to you? It's said to cure fever, headache, cramps, and pain in general. Do we really understand how it works? If anyone says they understand how the active and passive alkaloids interact, you tell them that Eddie Ayensu says they're lying." Certainly in many cases an extract of the whole plant produces a different, often less toxic, reaction than the active ingredient in isolation. One plant that is used in folk medicine in Ghana, *Cryptolepis sanguinolenta,* is, as laboratory tests confirm, a broad-spectrum antibiotic. Yet scientists have not been able to identify any antibiotic action in the individual compounds that they have isolated from the plant.

"In the Third World," Ayensu goes on, "we don't have the U.S. Food and Drug Administration. People are the guinea pigs and have been for millennia. There are more doctors in the three-hundred-acre National Institutes of Health complex near Washington, D.C., than in all of Africa. Eighty percent of people in the world will rely on herbal medicine from the day they are born until the day they die. And I don't think this should change. But we can refine uses and doses of medicinal plants. Some people are trying to find ways of turning herbal medicines into Western-type drugs. This is important, but it's more important still to try to look at herbal medicines as they are used. It is easy now to assay a plant for alkaloids. And you can now tell if there are toxins in the plant, so you know how strong the dose can be."

. . .

Forest tribes, who have lived there for hundreds or thousands of years, know far more about the plants of the forest and how to use them than any outsiders. Many plants are safe and effective only after complex treatments that have evolved over many years of experiment. Epena is an important narcotic with great medical potential. For centuries South American Indian tribes have used it, but Western scientists discovered only in 1938 that it comes from the genus *Virola.* Fifteen more years elapsed before its preparation and use by Indian medicine men was described in detail.

Epena is prepared in different ways, depending on the species of *Virola* used and whether it is to be used in religious ceremonies or as medicine. Several tribes in eastern Colombia begin by stripping the bark from the trees in the early morning (when the resin is most abundant). The medicine man scrapes the narcotic resin from the inner bark and kneads the scrapings in cold water for twenty minutes. The liquid is then filtered, boiled to a syrup, and dried. The dried substance is ground up, mixed with the ashes of the bark of a wild cocoa tree, and taken as snuff.

The Witoto, who live in the Colombian Amazon, scrape the precious resin into a gourd and knead it to release a liquid that is then boiled, possibly to inactivate enzymes that might destroy the active principles. The liquid is then simmered and reduced to a thick consistency and rolled into pellets to be eaten. If the pellets are to be kept longer than two months, they are coated with a preservative "salt," which is made by burning a number of specified plants and filtering water through their ashes. The water is collected and boiled until only a grayish residue, the salt, remains.

The way epena is made, as well as the rituals that control its use, enable some Indian groups to take what Richard Evans Schultes describes as "frighteningly excessive amounts" with no long-term ill effects. According to Schultes, who spent many years living and collecting plants among the Indians of Colombia, they

> exhibit uncanny knowledge of different "kinds" [of *Virola* trees]—which to a botanist appear to be indistinguishable as to species. Before stripping the bark from a trunk they are able to predict how long the exudate will take to turn red, whether it will be mild or peppery to the tongue when tasted, how long it will retain its potency when made into snuff, and many other hidden characteristics. Whether these differences be due to age of the tree, season of the year, ecological situations, condition of flowering or fruiting, or other environmental or physiological factors it is at present impossible to say.

. . .

The human race today gets more than half its protein and calories from just three species of grass—wheat, rice, and maize. Another ten crops—soybeans, barley, oats, peanuts, potatoes, millet, sorghum, rye, sweet potatoes, and coconuts—provide all but 10 percent of the rest. Half of these, including rice, have wild ancestors in tropical rainforests. Other important crops that originate in rainforests are coffee, bananas, eggplants, lemons, oranges, tea, cacao, cashews, cassava, peanuts, pineapples, guavas, and papayas.

In the old days farmers fought pests and diseases, as well as adverse soil and weather conditions, by cultivating many different varieties of a crop. The people in just one area of the island of New Britain, part of Papua New Guinea, have developed two hundred cultivars of their staple crop, taro, each with different characteristics. Variability is just as important to the modern agribusinessman, but he gets it by changing strains regularly. Modern crop varieties, planted in monocultures, offer a clear field to predators and pests. They must be replaced every five, ten, or fifteen years as pests and diseases adapt to them. Crop breeders need genes from wild plants and from the primitive crops of forest farmers to fortify modern varieties. Explorers have recently collected rice plants from subsistence farmers in the foothills of the Himalayas that have turned out to be resistant to many major pests and diseases including blast, bacterial blight, tungro virus, gall midge, and stem borer. But these wild species and ancient varieties are disappearing as forests are cut down and farmers switch to modern varieties. These, the so-called high-yielding varieties, are really high-response varieties that require high levels of expensive and sophisticated inputs—from chemical fertilizers to irrigation.

Ironically the success of the new breeds is undermining the position of the varieties on which their breeding programs depend. In the late 1960s researchers at the International Rice Research Institute in the Philippines screened about ten thousand varieties of rice, looking without success for a gene that was resistant to grassy stunt virus, the only major rice disease for which there was no protection. Finally in desperation they turned to their collection of wild rices.

They found one species, *Oriza nivara*, with resistance to the virus. Not all seeds, even of the same species, are genetically identical, and of the hundreds of seeds of *O. nivara* that the institute had, only two seeds from one particular location in the Indian state of Madhya Pradesh possessed the resistant gene. Collectors returned to gather more seeds from that area, but none had the resistant gene. In the fifteen years since that discovery, no one has found another source of resistance to grassy stunt virus. Every modern rice plant has a gene derived from one of those two original seeds.

Cocoa plantations throughout the world rely on genes from wild and semi-wild Amazonian cocoas. The Amazonian varieties have boosted yields and provided resistance to several diseases. Similarly, wild rubber trees from Peru produce more latex than the cultivated varieties, and breeders are now working to introduce this trait into Malaysian rubber plantations, along with resistance to South American leaf blight found in some Brazilian species. Wild tomatoes from Ecuador, Chile, and Peru have improved the color of cultivated tomatoes and given them more vitamin C and a higher soluble-solids content, an improvement worth $5 million a year to the tomato-processing industry. Wild West African coffees are providing coffee grown on Colombian plantations with genes that are resistant to rust. Breeders recently crossed several wild peanuts from Amazonia with cultivated peanuts to produce a strain that is resistant to leafspot. That improvement is estimated by the International Crops Research Institute to be worth $500 million a year. In 1975 international trade in just nine rainforest spices—pepper, ginger, cloves, cinnamon, cassia, mace, nutmeg, allspice, and cardamom—equaled $144 million. The Caribbean pine, native to the rainforests of Central America, a mainstay of many tropical plantations around the world, is one of the most valuable rainforest plants. Its growth rate and suitability for different areas could be much improved by breeding programs using its wild relatives. Unfortunately, only a few sizable Central American rainforests are left, and their survival is uncertain.

Ninety-eight percent of crop production in the United States is based on non-native species. Some experts say that the American economy is more dependent on foreign sources of germ plasm than it is on foreign sources of oil. Not long ago in Malaysia a botanist picked up a fruit that had fallen in his path. It was a rare find: a new species of citrus. Citrus is a multimillion-dollar crop in the United States, and very little is known about its wild relatives. Experts are enthusiastic about the new find, which they believe may have useful traits for cross-breeding. One possibility is that they will be able to introduce the new species' tolerance of high rainfall into other citrus varieties—an improvement that may mean millions of dollars to U.S. growers.

Forest people eat many plants (an estimated four thousand in Indonesia alone) that are unknown to outsiders. In 1970 almost no one outside Southeast Asia had heard of the winged bean, although New Guinea tribesmen had been cultivating it for many generations. Then in 1974, the U.S. National Academy of Sciences "discovered" the fast-growing, protein- and vitamin-rich bean and recommended it as an important food crop in the humid tropics. People in seventy countries now grow and eat winged beans.

The U.S. National Academy of Sciences has said that many plants that are now collected from the wild or grown as subsistence crops could, with good breeding and appropriate techniques, be important crops in the tropics. Among the plants the academy says should be grown more widely are cocoyams and taro, two nutritious potatolike tubers; wax gourds, which produce three or four crops a year of a mild, juicy melonlike vegetable that can be stored for up to a year even in the humid tropics; and six fruits, including the mangosteen, which many people believe is the world's best-tasting fruit. Generations of Indians in Paraguay have used the leaves of the *Stevia* plant as a sweetener. Now Japanese researchers have analyzed it and found a chemical that is, according to the Japanese Ministry of Health, calorie-free, harmless to humans, and three hundred times sweeter than sugar.

Insects and animals of the rainforest also contribute to modern agriculture. Each year insects do $2 billion worth of damage to U.S. crops, in spite of the fact that farmers spend billions of dollars annually on pesticides. The problem, according to the U.S. Department of Agriculture, is that about half of the country's five hundred major insect pests are now resistant to insecticides. "Biocides induce resistance," according to tropical ecologist Robert Goodland of the World Bank. "Every chemical pesticide selects for its own failure." One solution, called biological pest control, is to keep pest numbers down by bringing in other insects that are predators or parasites on crop pests. The USDA says that importing counterpests for biological control programs returns $30 for every $1 invested. Florida citrus growers have saved $30 million a year by importing three types of parasitic wasps to control pests—a once-only expense of $35,000. Hawaiian sugarcane plantations were suffering annual losses of from $750,000 to $1 million to the sugarcane beetle borer until 1910, when a parasite on the beetle, the tachinid fly, was found in the rainforest of New Guinea and brought to Hawaii.

The kouprey, a shy forest ox from which domesticated cattle are descended, is the most recent mammal to be discovered by science. There are images of the kouprey in prehistoric caves in Indochina and on the walls of Cambodia's ancient temple at Angkor Wat, but Western biologists first saw the living animal in 1937, when one arrived at the Paris Zoo in a job-lot shipment of animals from Saigon. Male kouprey are six feet across at the shoulder, with enormous dewlaps and almost the longest horns of all cattle. No one has seen a female close up, but they have beautifully curving horns. Both sexes are said to move with the grace of antelopes.

As a wild ancestor of cattle, the kouprey could be invaluable in breeding schemes, especially since it is reputed to be resistant to the widespread cattle disease rinderpest. Kouprey have always been rare —only five hundred were estimated to be alive in 1951. The last confirmed sighting of one was in 1949. The next thirty years were years of war and defoliation for Indochina, and the kouprey vanished. In 1972 a team of wildlife experts went to Cambodia to try and bring some kouprey back to safety. "We had good equipment, a trained team, and knew where to locate kouprey," said Pierre Pfeffer, one of the group, "but we had to wait for the monsoons to end." In the meantime the war escalated, forcing the abandonment of the project. Then, in the summer of 1982, a small herd of kouprey consisting of two cows, two calves, and a bull appeared in northeastern Thailand, on the borders of Laos and Cambodia. Immediately a team set out from the Royal Thai Forest Department to "rescue" the animals from the minefield-ridden area, the scene of frequent border clashes, and bring them back to a captive breeding station. So far they have not found them.

. . .

In many countries "wild" food—from hunting, gathering, or fishing—is important to the population as a whole, not only the forest dwellers. In Papua New Guinea people get 60 percent of their animal protein supply from wild animals and fish. Rural people in Honduras get 80 percent of their meat from hunting in the forest. In Benin, the Congo, Ghana, Togo, Zaire, Bangladesh, Indonesia, Malaysia, the Philippines, Thailand, and Vietnam more than half of the average daily animal protein supply for the country as a whole comes from the wild. To the people who actually live in the forest, wild game and fish are even more important.

Fish is the major source of protein for many inhabitants of the forest. Fish farming has great potential as a source of local protein and export income. But no tropical rivers or lakes are managed intensively in a manner comparable with the fishing areas of the developed countries. This is partly because our understanding of tropical rivers is at such a low level. According to Peter Raven, when we try to establish fisheries or otherwise manage tropical rivers, "we are attempting to manipulate a system in which only somewhat over half of the elements have ever been registered, much less understood." Few of the thousands of rainforest fish are known to any but experienced forest fishing tribes, but if even a small fraction of the available species became commercially acceptable, the economic potential would be enormous. The pescado blanco, a fish found only in one Mexican lake, was on the verge of extinction not long ago due to overfishing, habitat destruction, and competition from foreign fish that had been introduced into the lake. Now it is stocked in many reservoirs, and a thirty-six-acre fish farm is being built for it. The swamp forests of Cambodia's Tonle Sap region are ten times as productive as the best Atlantic fishing grounds.

. . .

The wild harvest of the forest is free. Its fruits, spices, game, fish, medicines, building materials, and firewood can supplement or even supplant farm crops and manufactured goods. At the same time, forest products can provide a cash income to professional collectors and forest dwellers who gather rattans, resins, latex, flavorings, perfumes, shellac, essential oils, gums, disinfectants, insecticides, fibers, glues, weed killers, dye, tannins, fats, and waxes from the forest.

Fifteen rainforest countries earn more than $20 million a year from forest product exports, according to the Food and Agriculture Organization of the United Nations. In Southeast Asia rattan is the second most important forest product after timber, worth $1.2 billion in 1977. India's wild forest harvest of perfumes, flavorings, resins, rattans, essential oils, and drugs is worth $125 million every year. In the Amazon a good stand of brazilnut trees is more profitable and provides more jobs than an equivalent area of pasture and does not require an initial investment to clear, fence, and weed the land. When brazilnut trees are felled, the people who once supplemented their income by collecting and selling the nuts are forced to clear more forest so that they can grow more food to make up for their lost income.

Plants and animals are sometimes simple tools of survival, as are the army ants whose pincers Amazonian Indians use to stitch up their wounds. Or they may become high-technology industrial products. An Asian plant, commonly called guar, has a host of industrial applications, from making water slide through fire hoses more easily, to thickening ice cream. Unilever is developing a "sweet protein," sixteen hundred times sweeter than sugar, from a West African plant, *Thaumatococcus daniellii,* to be used in industrial fermentation.

Many rainforest plants are rich in oil. For many years collectors tapped copaiba trees in the upper reaches of the Amazon for "liquid balsam" and floated it downriver in hollow trees; it was then shipped to Europe and North America for use in medicine and perfumery. In 1979 Melvin Calvin, the Nobel Prize–winning biochemist at the University of California at Berkeley, discovered that copaiba's liquid balsam is almost identical to diesel fuel. Tapping the trunk can yield four gallons in two hours, and the sap, poured straight into a fuel tank, can power a diesel truck. Related species, members of the legume family, grow in the forests of Latin America and Africa. Some are now being grown in an experimental plantation on Okinawa.

Another Amazonian tree, *Jessenia polycarpa,* produces an oil very much like olive oil that can be bought only in peasant markets in Bogotá. Local people drink the milky residue that remains after the oil is extracted from the seed; the seed itself is also edible. Babassu palms produce fruit that looks, tastes, and smells like coconut but contains 72 percent more oil than coconut. The oil can be used in producing fibers, soap, detergent, starch, and many kinds of food. After the oil is extracted, the leftover seed cake makes a high-protein animal feed. The husk can be burned directly as fuel or converted to charcoal. There are thirty million acres of babassu forest in Brazil, much of it threatened because it grows in areas that are scheduled for large development projects.

Philippine tribal people have long fueled their wick lamps with oil squeezed from the fruit of the petroleum nut tree. The oil is high octane; during the Second World War the Japanese used it as fuel for their tanks. It could be an excellent source of household fuel, since six trees grown in a home garden can produce seventy gallons of oil a year; and because it does not require a great deal of rain, it could be useful throughout the dry tropics as well as in rainforests.

. . .

Although national laws and the Convention on International Trade in Endangered Species ban trade in wildcats because most species are now on the verge of extinction, many such animal furs are available on the black market. A few years ago in Munich ocelot coats (each made of ten skins) were selling for $40,000. The prices at source are substantially lower. In Amazonia, where, although it is illegal, professional hunters still kill large numbers of cats, a jaguar skin fetches around $40, an ocelot skin slightly less. The pet trade is the main market for rainforest birds, although the vast majority suffocate to death in the miserable cages in which they are smuggled. A hyacinth macaw from the forests of South America costs the jaded bird fancier in the United States $8,000. According to Harald Sioli, one of the world's greatest experts on the Amazon rainforest, for years five million cayman skins went to the tanneries in Manaus each year. As a result, the cayman has "now almost completely been transformed into belts, purses, and billfolds."

In 1913 Theodore Roosevelt led a scientific collecting expedition through Brazil. Actually the great Brazilian explorer and champion of the Indians, Col. Candido Rondon, led the expedition and shielded Roosevelt from all discomfort, although this was not reflected in the book Roosevelt later wrote about the trip. That book, *Through the*

Brazilian Wilderness, did as much as anything else to create in the public mind an image of the exotic beauty and danger of "the jungle," exhibiting his tangled view of hunting and conservation, sport, food, and science.

> Several times we saw the giant ant-bear. Kermit shot one because the naturalists eagerly wished for a second specimen; afterward we were relieved of all necessity to molest the strange, out-of-date creature.
>
> At one camp Cherrie collected a dozen perching birds; Miller a beautiful little rail; and Kermit, with the small Luger belt-rifle, a handsome curassow, nearly as big as a turkey, out of which, after it had been skinned, the cook made a delicious canja, the thick Brazilian soup. . . . All these birds were new to the collection—no naturalists had previously worked in this region—so that the afternoon's work represented nine species new to the collection, six new genera, and a most excellent soup.

Despite his lust for the hunt, or—avid hunters might say—because of it, Roosevelt had a conservationist streak, at a time when few people were addressing the question. "There is every reason why the good people of South America should waken, as we of North America, very late in the day, are beginning to waken . . . to the duty of preserving from impoverishment and extinction the wild life which is an asset of such interest and value in our several lands." The question is how this can be done. One answer is by establishing national parks or other protected areas.

About 2 percent of the world's rainforests have been declared nature reserves or national parks. The vast majority of these are completely unprotected, and some are leased for logging or other disturbing activities. Brazil has one of the most impressive systems of protected areas of all rainforest countries. In 1979 there was only one national park in the Amazon rainforest. Now there are five, plus three biological reserves and five ecological stations, set aside for scientific study. Altogether they cover 40,000 square miles, 2 percent of the area termed Amazônia Legal, and the government says it will eventually bring another 30,000 square miles under protection. But only one of the national parks and one of the biological reserves have a director, guards, rangers, and the money to carry out essential works, such as marking the boundaries. The rest are what is known in the trade as paper parks. "All our parks and reserves," Brazil's campaigning environmentalist and perennial thorn in the side of the government, José Lutzenberger, told me, "are in a shambles; there is not a single one that is protected." Government officials acknowledge the problem. In the rush to exploit Amazonia's mineral reserves and waterpower, nature reserves often get short shrift. Cachoeira Porteira, a dam on the Trombetas River, will flood part of the Trombetas biological reserve, the only one that is *not* a paper park. The Grande Carajás Project overlaps Amazonia National Park, on the Tapajos River, Maria Tereza Padua, until recently Brazil's director of national parks, pointed out to me. It is, as she said, one thing to create a national park, to negotiate the minefield of objections from other government agencies, and quite another thing to make it work. "They say yes to the creation of a new protected area and afterwards they then try to put in roads, mines, and so forth. They're not allowed to, but it's hard to stop them."

The problem is by no means confined to Brazil. Indonesia is another country with an impressive system of paper parks. "We have sixteen parks, but we don't have sixteen biologists who can really manage them from a biological point of view," Indonesia's environment minister recently confessed. The Meru Betiri Game Reserve in eastern Java covers 125,000 acres. It is thought to be the last place on earth where Javan tigers survive; in fact, the entire wild population is down to one pair, and even their existence is in doubt. To protect this very endangered, and possibly already extinct, species, the government of Indonesia decided in 1979 to enlarge and upgrade the reserve to a national park. The new boundaries enclosed two large tree crop plantations and the homes of the workers on those plantations. The plantations were abandoned, and all the people living within the new boundaries were moved elsewhere.

Such disruptive measures in the name of wildlife are opposed by many ecologists, who point out that besides giving conservation a bad name, they often fail because they are isolated gestures rather than part of a long-term management plan. "Why," asked Kuswarta Kartawinata, the director of the Bogor Herbarium and one of Indonesia's leading environmentalists, "is a pair of tigers more important than an established plantation and the people who live within it?" It is a sad fact that severe conservation crackdowns are usually at the expense of the poor; seldom is big business similarly affected. East Kutai Nature Reserve, also in Indonesia, has not been subjected to the strict controls that forced the people of Meru Betiri to leave their homes. The 750,000-acre reserve is one of only two lowland reserves in East Kalimantan. In 1968, 250,000 acres were cut out of the reserve and given over to oil drilling and logging. Later another 150,000 acres inside the reserve were leased to a logging company. To compensate for that, the excised 250,000 acres, now logged over and occupied by oil drills and farmers, were returned to the reserve. Desperate to save the remaining forest, the Directorate of National Conservation obtained permission to turn the reserve into a national park, which has more status and merits greater protection. Logging continues there.

In less than thirty years the Ivory Coast has lost nearly 90 percent of its forest. The last remaining virgin forest in the country, and in the whole of West Africa, is the 1,400-square-mile Tai National Park. The forest has more than 150 plant species that grow nowhere else, and a rich wildlife, including pygmy hippopotami and elephants. The Ivory Coast's timber industry will collapse around 1988, when the last forests are exhausted—unless the government allows loggers into the park for a few more years of harvesting. The southwest corner of the country, where Tai is, was largely unpopulated until the mid-1960s, when the government decided to develop the region. The complex includes the city of San Pedro, a major seaport built for the export of timber to Western Europe, a paper and pulp mill, a hydroelectric dam, a railway, peasant colonization areas, and various plantations and food-processing factories. So far relatively few of the farmers who gravitated to the area have settled in the park, though as the land outside is colonized, pressure on the forest increases. All this, plus large-scale organized poaching and illegal logging, has seriously disrupted the lives of the local tribal people and of the wildlife of the forest.

The very survival of the rainforest, say scientists working there, is at stake. Already one quarter of the park has been destroyed. "The situation is reminiscent of the U.S.A.'s Wild West, with everyone seizing as much land as they can, while they can," remarked one. Poaching has reduced the elephant population to fewer than one thousand; one truck that was intercepted had eighty-two tusks. Until recently twenty or more huge trucks loaded with logs were leaving the park daily on a road illegally built by a logging company.

Things may be looking up for the Tai as a result of work by the International Union for the Conservation of Nature (IUCN) and diplomatic moves by Prince Bernhard, formerly president of IUCN's sister organization, the World Wildlife Fund. Largely as a result of discussions between golfing partners Bernhard and Felix Houphouet-Boigny, the president of the Ivory Coast, government ministers are under order to take the park seriously. This means allocating the money and manpower necessary to protect it, clamping down on local businessmen whose connections with local politicians and judges have allowed them to flout the law by poaching and logging, stopping government departments from carrying out damaging activities in or near the park, and providing alternatives for landless people who would otherwise settle in the forest. Right now it is too early to say whether the president's order will have any effect.

Not all is bleak. There are a few national park success stories. Costa Rica is one of the growing handful of countries (including Indonesia, where there are more than four hundred citizen nature groups, and Brazil) in which conservation has much popular support. It has an excellent system of parks and protected areas. Groups of birdwatchers, college science students, and other nature lovers have created a highly profitable "educational tourism" industry that is an important economic factor in Costa Rica and a strong argument in favor of protecting protected areas.

One of the most successful attempts to get national parks to pay for themselves through scientific tourism is the Mountain Gorilla Project in Rwanda. There, in the most densely populated country in Africa, several international conservation groups, with the blessing of the government, may have saved the Volcans National Park, which was slated to be destroyed as part of a cattle ranching project, by making it profitable. Before the project began, the thirty-thousand-acre park, established in 1930 as a gorilla sanctuary, was being steadily diminished. In 1962 twenty-two thousand acres were removed from the park and given over to the planting of pyrethrum, a daisylike flower that produces a natural insecticide, for export to Europe.

The Virunga Mountains, of which the park is a part, are home to the last surviving mountain gorillas in the world. Europeans discovered mountain gorillas only in 1901; today there are only 500 left in the world. They are on the verge of extinction. The park's population has declined to 240 from around 450 in 1960. Hunting is one of the main threats to their survival. Poachers take the gentle creatures, not for meat, but in order to sell their heads and hands as souvenirs —the hands are used as ashtrays—and to sell the young animals to collectors. Several gorillas in the park today have lost their hands in snares laid out for other animals.

In 1979 the Mountain Gorilla project set out to do three things: provide equipment and training to enable the park guards to reduce poaching, inform Rwandans about the benefits of having the national park, and make tourism pay for the first two activities. Once a day a small group of tourists, and guides, climbs as high as nine thousand feet into the mountains looking for gorillas. The gorillas allow their visitors to come within two or three yards of them. "Some of the groups are very curious—they come right up to you," said Conrad Aveling, who with his wife, Rosalind, worked for several years on the project. "We try and discourage this because contact brings the danger of transmitting disease to them. . . . But gorillas are such peaceable and trusting animals that it is hard to avoid contact. I have a picture of the leader of one group of gorillas, who lost a hand in a wire snare, touching, very gently with his stump, the hat of the Rwandan guide. Somehow it is a very touching scene. They are extraordinary animals." The visit costs about $40, and since a maximum of fourteen people can go each day, the trips are often booked three months in advance. In 1981 the national park broke even; in 1982 it made "quite a good profit." Revenues from tourism have tripled since the project began, and it is now Rwanda's fourth largest source of foreign exchange. Every secondary school in the country has had a slide show and talk about the gorillas and the role of the park in protecting water supplies. Groups of Rwandan schoolchildren visit the park; most, seeing their own native wildlife for the first time, are overwhelmed, according to the Avelings. The success of the project gives strength to those who are now arguing against a Ministry of Agriculture plan to set aside twelve thousand acres of the park for cattle-raising.

Few countries or parks are as lucky as Rwanda in being able to keep tourism under control and make it profitable. International aid agencies pay lip service to conservation but put very little money at its service. According to Marc Dourojeanni, the former director of Peru's National Park Service, there is not a single conservation project in Peru in which Peru is not paying 80 percent of the costs.

The notion that rainforests are global treasures that should be preserved for the good of all mankind does not go down well in many rainforest countries. Why, they ask, should we have to make up for the fact that Europe and North America destroyed their own primeval forests centuries ago? Nor do they much like being told to live in harmony with nature by countries that appear to be enjoying the sweet fruits of centuries of assaults upon the natural world. This attitude was clearly demonstrated a few years ago at a meeting organized by several United Nations agencies. The idea was to draw up what was called a Plan of Action for the Wise Management of Tropical Forests. Denmark, France, West Germany, Japan, the Netherlands, Norway, Britain, and the United States all turned out for the meeting. Brazil, Zaire, Colombia, and Venezuela did not.

Trying to solve the twin problems of sovereignty and money, Ira Rubinoff of the Smithsonian Institution has proposed a worldwide system of protected reserves covering 10 percent of all rainforests (a total of 375,000 square miles). The program would be voluntary, and countries would be compensated for setting aside the reserves. Rubinoff estimates the plan would cost $3 billion a year (about what the world spends on armaments every thirty-six hours), to be raised by a levy on all countries with a per capita income greater than $1,500 a year. If rainforests are part of the earth's heritage, argues Rubinoff and a number of rainforest countries where his proposal has been favorably received, the international community should be prepared to pay for their preservation.

"The South," Emil Salim said to me, "is asked to conserve genetic resources, while the North is consuming things that force us to destroy our genetic resources. Who is going to pay for this protection? We should strive for an equitable share in the benefits of our genetic resources, and an equitable share in the costs of safeguarding them."

Genetic resources—the very phrase implies that plants and animals exist to serve mankind. The biologist Michael Soulé warns against trying to "sell" conservation by promising new industries and wonder drugs. Tom Lovejoy, a tropical ecologist and vice-president of the World Wildlife Fund, agrees that not everything can be priced. "Sometimes I'm asked: What difference does it make whether there's one species or ten species less in some tropical forest? My answer is that I probably can't tell them what an individual species is doing, but I can say that it's making an incremental contribution to maintaining the stability of global chemistry and climate.

Richard Spruce addressed the question in a letter to a friend in 1873. "It is true that the Hepaticae [liverworts] have hardly as yet yielded any substance to man capable of stupefying him, or of forcing his stomach to empty its contents, nor are they good for food; but if man cannot torture them to his uses or abuses, they are infinitely useful where God has placed them, as I hope to live to show; and they are, at the least, useful to, and beautiful in, themselves—surely the primary motive for every individual existence."

14

The Flooded Forest

In the nineteenth century rubber made Manaus, a steamy settlement two thousand miles up the Amazon River, one of the wealthiest cities in the world. It was one of the first to have electric trams, beating Boston and Manchester in that race. At the turn of the century Manaus supported eight daily newspapers and boasted a telephone network.

The opera house, the Teatro Amazonas, completed in 1896 at a cost of £400,000, is a monument to the pretensions of old Manaus. Its Byzantine golden dome perches over a "Renaissance" facade. Inside, instead of banks of folding seats, are rows of free-standing Queen Anne–style chairs upholstered in red velvet. Four tiers of boxes, draped in red velvet and fronted with golden balconies, circle the room. A wonderfully florid mural, in which the arts and muses cavort with assorted cupids and baby angels, covers the ceiling. A *trompe l'oeil* bust of the man who built the Teatro, a Mark Twain look-alike named Eduardo Ribeiro, looks down over the stage. Ornately gilded columns supporting the first balcony are topped with scrolls inscribed with the names of artists apparently claimed by Manaus as her spiritual sons: Goethe, Mozart, Beethoven, Verdi, Racine. The effect is of an oversize Italian biscuit tin.

A local singer-songwriter who goes by the single name Adelson appeared at the opera house while I was in Manaus. Adelson's backup band included several unweaned babies and their mothers and a violinist named Nelson Eddy. The audience was composed almost exclusively of people in their teens and early twenties, and I was braced for an evening of loud Brazilian rock. But Adelson's songs turned out to be gentle and witty celebrations of various local phenomena. One was a serenade to the tukunare, one of the most delicious Amazonian fishes. Another sang of the creaking sound made by a hammock hook as it swings back and forth with its load, and a third of Tarzan, the white man's King of the Jungle.

When I realized that the finale, which brought on several encores, was about deforestation, I began scribbling notes. The song told of so many trees being cut down that the red earth of the forest ran into the white waters of the Solimões, the local name for the Amazon. "As far as the Solimões goes, white and red don't mix. Don't kill the forest. The green virgin deserves consideration." My rough translation doesn't do justice to Adelson's words. But the enthusiastic reception the song received owed more, I fear, to its appealing rhythm than to its ecological sentiments.

At Manaus the Solimões is joined from the north by its main tributary, the Rio Negro. The Solimões rises nine hundred miles away in Peru and flows through the Andes down into the Amazonian lowlands. By the time it reaches Manaus, its waters are murky with silt washed down from the relatively young (100 million years old) Andean soils. The river supports huge communities of plants, some of which are not even anchored to the riverbed but float freely, supported by the nutrient-rich waters. By contrast, the waters of the Rio Negro and other so-called blackwater and clearwater rivers that arise in the ancient and more weathered soils of the Guianan and Brazilian Highlands are nutrient-poor and clear, the color of weak tea. The difference is nowhere more clear than at Manaus, where the Rio Negro and the Amazon come together. The current of each river is so strong, and the differences in flow and density so great, that the yellow, foamy Amazon and the clear, smooth Negro flow side by side, not mingling, for many miles.

It is pleasant to travel on the Rio Negro. There are hardly any insects (because of the paucity of nutrients), and the water is placid and silky to swim in. I traveled some seventy miles up the river with Michael Goulding, an American scientist working in Brazil, to a floating ecological research station built by the Brazilian government in the center of a breathtaking archipelago called the Anavilhanas.

The river here is twenty miles wide from bank to bank. At one point, when we were several miles away from the east bank, I asked why the forest alongside the river was so short. The trees looked stunted. Was it, I wondered, something to do with soil conditions? Goulding looked amused and asked the boatman to take us nearer. As we approached the bank, I could see that the reason the trees looked so short was that they were largely under water; only the top halves showed above the high water level. The boatman, to whom Goulding turned, guessed that we were floating thirty or forty feet above ground level.

Forty thousand square miles of Amazonia are regularly flooded like this—a huge area, though only 2 percent of the whole Amazon Basin. In some places the floods extend sixty miles on either side of the river. The trees and shrubs of these forests are adapted to living under water for as many as ten months of the year. Seeds can germinate in the few months of relative dryness; seedlings can live completely under water for long periods and emerge to spurt up during the short growing season. Scientists don't yet know how plants survive this degree of flooding, although some trees appear to have rootlike structures high above the flood line, where more oxygen is available.

Canoeing silently through the flooded forest is magical, like being in another, charmed world. Tropical forests are as a rule taller than temperate ones, and most of the life of the forest goes on high up in the canopy, where it is hidden from humans. We were privileged simply to drift through the canopy, silently and effortlessly. Two pairs of blue and yellow macaws flew close overhead, showing themselves to us like a gift. At the forest's edge, dolphins leapt out of the water. A troupe of spider monkeys scampered wildly among the branches of a fig tree; the pink and white of an orchid stood out among its shadows. Down below a cayman, invisible in the forest shade, betrayed its presence only by the thud of its tail as it jerked against the water.

The beauty and life of this river belie a kind of poverty. Like the rich rainforest that thrives on poor soil, blackwater and clearwater rivers are so lacking in nutrients that they should not be able to support the mass of fish and animal life that they do. Goulding has discovered that in this river and others like it, the fish depend upon the trees of the forest. Fruit that falls into the river forms the major part of the diet of at least fifty species of fish here, including a number of species of vegetarian piranha fish; another two hundred fish species eat the fruit eaters. Many fish have large molars and specialized jaws for cracking hard seeds. They eat well in the forest and build up stores of fat. In fact many species appear not to eat at all once the floodwaters have receded and they can no longer reach the fruits of the forest. After one dry-season fishing trip on the Rio Madeira, Goulding analyzed 167 fish and found that not one had significant amounts of food in its stomach. Among the fish that depend on the flooded forest are many of the largest and most commercially important species, such as the tambaqui, which appears to depend largely upon the seeds of wild rubber trees. Goulding estimates that about 75 percent of all the fish sold in Manaus, the largest market in western Amazonia, depend ultimately upon the flooded forest.

All the rivers of Europe combined support fewer than 150 species of freshwater fish, but exploring near Manaus, the nineteenth-century zoologist Louis Agassiz found more than two hundred species in one lake alone, Lake Hyanuary, which is only about twice the size of a tennis court. According to Goulding there are at least seven hundred fish species in the Rio Negro alone. That is six or seven times as many species as in all of North America. No one knows how many species there are in the Amazon Basin as a whole, but five thousand (of which only three thousand have so far been described by scientists) is an often-quoted estimate—making the basin the world's most diverse collection of fish. The fish market at Manaus offers more than 150 different kinds of fish, several of which, to my taste buds at least, are to plaice and cod as mangoes are to Golden Delicious apples. Fish is the major source of protein in the Amazon. Goulding cites recent studies that show that "though poverty is at a distressing point [in Manaus], per capita daily protein intake was more than satisfactory and mostly because of a relatively cheap supply of fish from the region."

Ten years ago no one realized that the fate of Amazonia's fish might depend so closely on the fate of the forests. Now Goulding's findings are quoted by every government official as one of the strongest arguments against clearing the riverine forests. Where these forests have been cut down, researchers, and fishermen, have measured a serious drop in the local fish population. Between 1970 and 1975 the fish catch in Amazonian rivers fell by 25 percent due to deforestation of the fishes' breeding grounds. Some Amazonian floodplain forests are dying, even though they are not being logged, according to Brazilian ecologist José Lutzenberger. Deforestation upstream is altering the pattern of flooding to which the forests are adapted, and the resulting extremes of wet and dry are damaging them.

Some scientists say that if Amazonian rivers were properly managed for fishing, their production could dwarf that of the Mediterranean and the North Atlantic. Others, such as Goulding and José Lutzenberger, argue that the idea that the Amazon could feed the world is a dangerous illusion. "There is enough for a growing local population but not for large-scale export," says Lutzenberger. Already several species, including the pirarucu and the tambaqui, are nearing depletion.

Goulding, a born-and-bred Californian, fluent in what Doonesbury cartoonist Gary Trudeau calls mellowspeak, is a dead ringer for the blond policeman in the television series *Starsky and Hutch.* It is somewhat surprising to discover that he is also a natural historian in the tradition of the literate, scholarly explorers of the nineteenth century like Wallace and Spruce. He is a true field scientist, in love with nature rather than with his laboratory. He has a strong sense of history and an enthusiasm bordering on reverence for his predecessors in the field—the native peoples of the forest, as well as earlier explorers. "We're just discovering things that the Indians have known all along. The Indians know the stories of all these fish. They know what they do. All you have to do is look at the names they've given to the fish. Lots of the fish here are named after the trees whose fruits they eat. The tabebuia tree over there, with the big pink trumpet-shaped flowers—that has the same name as the male turtle that eats its seeds."

The Uanano Indians live on a tributary of the Rio Negro. They eat mostly fish, although they also grow a few crops. Their gardens, however, are never on the riverbank; they are always deep inside the forest. The river forest is guarded as a source of food for the fish. The

Uanano believe that the fish are looked after by "fish elders." The elders punish anyone who interferes with the supply of fruits to their fish, especially during the spawning period.

Spruce learned from the Indians over one hundred years ago that fish eat the fruits of the forest. He mentioned it in his journal: "The principal subsistence of fish in the Rio Negro is on the fruits of riparial trees. When the ripe drupes are dropping into the water they attract shoals of Uaracu [different species of *Leporinus*]. Then the fisherman stations his canoe at dawn in the mouth of some still flooded forest, overshaded by bushes of Uaracu-Tamacoari [the native Indian name of the tree], and with his arrows picks off the fish as they rise to snatch the floating fruits." Spruce's observation was overlooked by generations of scientists, who preferred to start their investigations from scratch.

Spruce was also much taken with the taste of fruit-eating fish. "It ought to be mentioned that the fish of the Negro, if much fewer, are some of them perhaps superior in flavour to any Amazon fish, whereof the Uaracu is an example, and the large Pirahyba is another, the latter being so luscious that it is difficult to know when one has had enough of it."

In the years Goulding has spent studying the rivers and fish of the Amazon he has learned a great deal from the fishermen and the riverbank peasant families. "The skillful caboclo [Amazonian peasant]," he has written, "has several tricks for capturing the tambaqui based on knowledge of its feeding habits. One device, called a gaponga, consists of a ball-bearing-like weight connected to the end of line and pole; the water is beaten in such a manner to imitate falling fruit, and when the fooled tambaqui surfaces, it is harpooned."

Tambaqui are so popular that they have been overfished, not by peasants catching their own food but by commercial fishermen using gill nets. These are walls of netting that are hung vertically in the water so that fish become entangled in them. There is no effective control of gill-net fishing in the Amazon Basin, and Goulding fears that some species may be fished out if its use continues to spread.

Another fruit-eating fish is the pirarucu. It is one of the largest freshwater fish in the world—often six feet long and over 170 pounds. "Next to a jaguar and a manatee," says Goulding, "a large pirarucu is the most prestigious animal that can be claimed by a caboclo." Pirarucu breathe air and so have to surface every ten or twenty minutes, which makes them very vulnerable targets. They have been intensively hunted in the Amazon for over a century. Dried pirarucu were the main rainy season protein food in the Amazonian interior, but due to overfishing they are now scarce and too expensive for most people.

Alone of the Amazonian fishes I encountered, the name of the pirarucu was familiar to me. I could not remember why. I assumed I had heard someone raving about how delicious it is. Months later, when I was back in London, I came across a slip of paper that reminded me of my first acquaintance with pirarucu. Several years ago, staying with friends in New York, I had been unable to resist buying them something I noticed in a sort of secondhand book-*cum*-junk shop. A "Natural Nail File," with an enthusiastic message from the distributors.

> What is it? Where does it come from? Your nailfile is actually one of the scales of a fish . . . yes a very large fish. The PIRARUCU fish is the second largest freshwater fish in the world, only exceeded by the sturgeon. He grows to a size of 6 to 12 feet long and may weigh hundreds of pounds. He swims in the Amazon river in South America, and has been used by the Indians there for hundreds of years. He is a basic food fish yielding succulent steaks; the tongue is dried in the sun and used as a wood rasp. The scales are used not only as nailfiles by the Indians, but also as jewelry and decorations. Nothing is wasted in Mother Nature's cycle. If engineers wanted to design a product to have the strength, longevity and functionality of NATURE'S NAIL FILE, they probably couldn't have come so close, so ENJOY!

15

That Ancient Flame

In 1933 the busiest airport in the world was a rough grass strip on a hillside deep in the uncharted rainforest of New Guinea. More planes landed and took off at Wau every day than at any other airport in the world. The reason was gold.

Europeans first heard of New Guinea in the early sixteenth century. But the various navigators who sighted land east of the East Indies, and later touched down on its coast, could not agree on whether it was an island or part of a vast antipodean continent that also included Tierra del Fuego. Some people believed that the new land was Ophir, the biblical land from which Solomon's ships brought back gold; others identified it with a legendary, gold-rich Pacific land thought to be well known to the Incas. The Spaniard Alvaro de Saavedro, the first European to see New Guinea, was moved in 1528 to call it Isla del Oro, island of gold. For the next 350 years and more, explorers circled the main island and its many satellites but scarcely penetrated the interior. Tempting but unsubstantiated tales of gold were offset by the very evident hazards of a swampy, densely forested coast.

When, in the late nineteenth century, European explorers finally began to brave the interior, they found it hard going. The flat land of the coasts soon gives way to the chain of rugged mountains that forms the backbone of the island, rising in many places to over ten thousand feet. The difficulties of traveling through rough and unknown terrain—dense forest, trackless hills, dangerous rapids, steep ravines—were compounded by the unpredictable reactions of the native people, who were and are as varied as the more than seven hundred languages they speak.

In 1828, with no idea of what they were claiming, the Dutch annexed the western half of New Guinea, now the Indonesian province of Irian Jaya. Britain and Germany both eyed the other half. In 1883 the *Age* and the *Argus,* two Melbourne newspapers that were campaigning for Britain to annex the eastern half of the island, sent expeditions to Papua. Local hostility and the rough terrain prevented either group from going more than forty miles inland. But there was a steady, if small, stream of explorers—naturalists, colonizers, missionaries, and adventurers—anxious to unravel the mysteries of the interior. One result of their work was that in 1885 Britain laid claim to the southeastern quarter of the island, which it called Papua. Another was the discovery of several gold deposits that, although modest, accounted for 54 percent of all Papua's export earnings between 1888 and 1898.

By the start of the Second World War, most of Papua had been mapped out. But exploration of the rest of the island got off to a slower start. The Germans, who took over the northeastern quarter in 1885 (loyally naming it Kaiser Wilhelmsland), had largely ignored the interior, concentrating instead on developing plantations in coastal areas. From about 1907, however, the German authorities began to allow, even encourage, prospectors to enter the territory in the hope that they would "open up" the interior and make it easier for plantation managers, who were running short of labor, to hire mountain people.

In 1909 two Germans, William Dammkohler and Rudolf Oldorp, found gold on a tributary of the Watut River, but what happened afterward hardly encouraged further exploration in the area. As the prospectors were sitting in their streamside camp one afternoon, a group of natives wielding bows and arrows attacked. The two Germans grabbed their rifles and shot out wildly, killing fifteen raiders. When the survivors retreated, Dammkohler lay dying, pierced by eleven arrows. Oldorp, badly wounded, managed to build a raft. He floated downstream for five days, losing his raft twice and nearly starving to death before reaching safety. The next year, having found a new partner, Oldorp set out again to look for the source of the gold he and Dammkohler had discovered. The day they arrived back in the area, a cyclone hit and both men were drowned.

Germany's reign over Kaiser Wilhelmsland ended during the First World War, when Australia assumed control. Afterward Australia administered the entire eastern half of the island as the Territory of Papua and New Guinea. Australia announced the change of government to the native population via posters written in pidgin English ("You look him new feller flag. You savvy him? He belonga British, he more better than other feller. . . . NO MORE 'UM KAISER, GOD SAVE 'UM KING").

Papua and New Guinea were returned to civil rule in 1921. The following year, after twenty years of prospecting in New Guinea, William "Sharkeye" Park found gold on Koranga Creek not far from where Dammkohler and Oldorp had hit it rich and died. Koranga Creek, a tributary of the Bulolo River, is only thirty miles from the east coast as the crow flies, but the terrain is so rough that it took eight or ten days to walk that distance. One early miner described part of the journey:

> We crossed a bridge over a stream. There were four stout bamboos for a footway and these and the hand-rail were secured by the vines. One at a time we crossed it, and as it swayed and dipped I tried not to think of the water and the rugged rocks beneath it. A fall meant death on the rocks below, and the bridge seemed very frail. . . . We ascended a gorge the walls of which were about 700 or 800 feet high, by a track going along one side. Here the stream met another one coming from the opposite direction, and the two joined in a cauldron of turmoil. . . . We came to another river, and proceeded along the slippery bed. Jumping from rock to rock, I feared that I should become a casualty, so dangerous was the going. At times the water was up to my arm-pits, the bed of this stream being very uneven. Swiftly-flowing currents almost carried one's feet away. The walls on both sides were very precipitous.

When he reached Koranga Creek, Sharkeye Park was in a valley about four thousand feet above sea level, surrounded by mountains that rose another four thousand, five thousand, or six thousand feet above him. Spurs ran down into the valley from the main ridges; rivulets (that become raging torrents after heavy rains) plummeted through deep gorges to the valley below. Towering over everything was Mount Kaindi, almost eight thousand feet high and shunned by the local people as cursed.

With its peak often hidden in clouds, Kaindi is very humid: Mosses, ferns, and other moisture-loving plants grow abundantly on trunks and branches throughout its many-layered forest. As Doris Booth, the only woman in the goldfields, later noted about the mountain, "there seemed to be only two seasons—the rainy and the wet." The uniform green of the forest is broken by the striking pink, white, or yellow blossoms of several rare varieties of rhododendron, some that grow nowhere else on earth. One of the strangest trees on Kaindi is *Phyllocladus,* a conifer whose stems look like leaves. It grows only in New Guinea and a few other Pacific islands. Two species of the southern beech, *Nothofagus,* growing in dense stands, with each magnificent, white-trunked tree up to five feet in diameter, dominate the upper reaches of Kaindi. *Nothofagus* grows only in New Guinea and a few islands nearby—and 3,500 miles away in a narrow band along the Pacific coast of South America.

The lower slopes and the valley floor are hotter and less humid: The cloud forest of the heights gives way to a slightly more open forest whose trees have larger trunks and crowns and are more widely spaced. Two species of tree dominate these forests in number and size. The hoop pine and the klinki pine, both members of the genus *Araucaria,* and both valuable timber trees, are sometimes more than twice as tall as the surrounding trees. Klinki pines, some of which are more than five hundred years old, grow only in Papua New Guinea. The branches of a klinki radiate in layers from its trunk, rather like a bottlebrush. *Araucaria* (a genus that includes the monkey puzzle tree) are among the most ancient and primitive plants in the world, closely related to trees that lived two hundred million years ago. They are among the very few pines that grow in lowland rainforests, and they grow more profusely in the Bulolo region than anywhere else. People throughout the island hold them sacred.

News of Park's strike took a long time to reach the coast, and until the middle of 1926 fewer than twenty miners were panning in the Bulolo Valley. Park himself was a rich man before too long; others were less lucky. Doris Booth and her husband, Charles, got only seventeen ounces of gold in their first year and a half.

Another miner who was having no luck was Bill Royal, an Australian who arrived at Koranga in mid-1924. With a partner, Dick Glasson, and six native carriers, Royal decided in January 1926 to make one last prospecting trip, up Edie Creek, which flowed down Mount Kaindi to the Bulolo. The climb up Kaindi was perilous. Because the mountain was taboo to the locals, there were no tracks. The ground was rocky and unstable: Small landslides were common. A thick carpet of moss concealed treacherous chasms. For much of the way the climbers had to pull themselves up sheer rock faces, grabbing at insecure vines and roots for support. It was a steep ascent, straight up for the first six thousand feet. One night, as they were making camp at seven thousand feet, Royal and one carrier went ahead to reconnoiter the next day's route. Almost at once they broke out of the forest and saw ahead of them a tableland leading to the summit and to the headwaters of the Edie. Royal scooped up a handful of wet gravel from the stream and sluiced it through his prospecting dish. As the water rolled the gravel aside, leaving chunks of gold in the bottom of the dish, Royal knew he was a rich man: He had discovered one of the richest deposits of alluvial gold ever found.

This time the word spread quickly, and by the middle of the year more than two hundred men were working the Edie, the Koranga, the Bulolo itself, or one of its many other small tributaries. Other prospectors poured into the Watut Valley, five miles and a mountain range to the west, where Dammkohler and Oldorp had made their strike. Each man had to bring in all his own supplies, which meant that one miner required as many as one hundred carriers to make the seventy-mile trip from the coast several times a year. Professional recruiters persuaded the reluctant coastal natives to make the arduous journey through territory belonging to hostile clans. Though Edie Creek was only seven thousand feet above sea level, the ten-day walk involved a total climb of twenty-nine thousand feet. By the end of 1926 recruiters were charging the miners £30 for each carrier supplied, exclusive of wages, which amounted to about £1 per man per trip. The law limited each carrier's load to fifty pounds, but more than one third of that consisted of his personal belongings and his food for the ten-day trip. Miners also relied heavily on native laborers to help work their claims.

The Bulolo-Watut region was not heavily populated, but there were villages scattered throughout the area, and the people there had a reputation for ferocity that made the coastal people, who had been pacified by German Lutheran missionaries, reluctant to go inland. Nonetheless, the first white men to enter the area, the miners with their queues of carriers, often came across deserted villages whose inhabitants were hiding in the bush, waiting for the strangers to pass. As the intruders became more numerous and their carriers, emboldened by strength of numbers, less timid, clashes became increasingly common and violent. Most of the casualties on both sides were natives, though brutal attacks on prospectors such as the one Dammkohler and Oldorp suffered were common enough for the authorities to declare six thousand square miles of Kukukuku territory off-limits to whites without special permission.

Many of the tribes in the region of Bulolo were said to be cannibals. Doris Booth wrote of stopping at one village near Koranga and being told that

> some of the boys had been kai kai-ing man [eating human flesh] and would I like to see a piece. I determined to see it and Usendran went off with one of the interpreters, and returned with a portion of a human foot! Proudly Usendran displayed the exhibit. I looked long enough to see that the skin was crackled on the foot, and that it looked like a piece of pork. Then I began to feel very queer, and I waved Usendran away, telling him to bury it.

Whether or not they practiced cannibalism, the people of the area were regarded as one of the most aggressive groups in mainland New Guinea. In 1936 an English anthropologist, Beatrice Blackwood, spent nine months at a Kukukuku village, where, though it was only half a day's walk from the goldmines, people were still to all intents and purposes living in the Stone Age. The Kukukuku are in reality not one tribe but a collection of clans, each with its own language or dialect, who in the past, at least, warred among themselves as frequently, if not as fiercely, as they did with outsiders. As they had no name for themselves, other tribes dubbed them the Kukukuku, a name the significance of which is obscure but which is nonetheless regarded as insulting by the people themselves. They prefer now to be called the Anga, meaning home, a name proposed by an anthropologist. Blackwood spent most of her time with the Manki subgroup, whose traditional lands include the area on the west bank of the Bulolo River.

Even as she wrote, Blackwood recognized that the culture of the Kukukuku was changing. "The government and the missions were trying to make the people come together in larger village groups, for convenience of administration, and to protect the weaker groups from being raided by the stronger ones." Many Manki were beginning to grow such vegetables as tomatoes, cabbage, and maize, which they would not eat themselves, in order to trade with the miners for metal tools. The village, Blackwood wrote, "was fast losing its old customs through contact with the miners and their indentured labourers on the Bulolo."

The Manki still hunted in the forest, using traps and bows and arrows, and gathered plants from it. Meat was not a large part of their diet, but they ate several small marsupials, including cuscus and wallaby; birds such as cockatoos, pigeons, and cassowaries; and a variety of other creatures—wild pigs, fish, tree rats, iguanas, lizards, snakes, and grubs. According to Blackwood, "Each plant, however insignificant, has its native name, which is known to almost everyone. . . . Nearly everything seems to be utilized by the natives for food, medicine, magic, personal adornment, or in connection with their arts and crafts. . . . My collection includes three species of edible bark."

The Manki were primarily shifting cultivators. They lived in small villages, collections of tiny hamlets, each with two or three houses belonging to one family group. The hamlets were scattered over the ridges and slopes of the mountains that separated the Bulolo and Watut valleys. The miners and the fiercer groups of Kukukuku had commandeered the gentler slopes and the flats. Around their settlements the Manki cleared the forest for gardens. Doris Booth had earlier noted that many of the gardens were on "the sides of mountains so steep that it seemed impossible that a footing could be obtained upon them." Families or clans held their gardens communally; the men cleared and fenced them, and the women planted, weeded, and harvested.

The Manki grew mostly sweet potatoes, but also taro, yams, bananas, maize, sugarcane, tobacco, and tomatoes. A family farmed its plot for a few years until yields began to decline, and then moved on to clear a new garden. After fifteen or twenty years, when the forest had recovered, they returned to the original plot and began the cycle again.

With stone tools—the local people had nothing else before the arrival of Europeans put steel implements into trade on the island—the Manki found it difficult to fell huge forest trees. To clear a garden plot, they generally waited until the forest was dry enough to burn and simply set fire to the undergrowth. The trees that did not burn were left standing. Blackwood recorded the Manki's way of cutting a large tree: "A platform is set up all around it at the desired height, usually about five feet, and on it sits a circle of men, each chipping away at his own bit. Someone will go off for a meal or a sleep and his place will be taken by another man, and so the work goes on for several days." With metal axes, however, it was easier to fell trees, and easier, therefore, once they had dried out, to set them alight.

Some gardens evidently did not lie fallow for long enough. They were cleared again before the forest was properly reestablished. If this happens often enough, the forest is permanently damaged and never recovers. In this area coarse grass, known locally as kunai (*Imperata cylindrica*, the dominant species, is identical to the elephant grass that covers much of Indonesia's degraded rainforests), creeps into the sunny areas where trees no longer shade the soil. Sooner or later, by accident or design, the grass, which burns easily, catches fire. The kunai grass then recolonizes the burned area; with its deep roots and tolerance of strong light, it is better adapted to do so than woody plants. Eventually the forest becomes a sterile grassland, with soils too depleted to allow farming.

The Manki followed the practice, common throughout New Guinea, of burning grassland frequently. They burned kunai to drive small animals into the open, where they could be hunted. They also set fires to signal from hill to hill. But, as Blackwood and many later observers have noted with incomprehension, Papuans often burn grassland just for the fun of seeing a fire, leaving it to burn itself out unless it threatens a settlement. One result is that even in the time of Blackwood's sojourn, the Watut and Bulolo valleys were "partly forested, partly covered with what appears from a distance to be soft turf, but is in reality coarse, intractable grass, often growing to a height of several feet." Though the valley was a patchwork of forest, gardens, and grasslands, however, Mount Kaindi, cursed and taboo, was still undisturbed forest.

Perhaps the happiest man to hear of Royal's strike at Edie Creek was C. J. Levien, who had been a district officer when Park made his find. Levien resigned his position in 1923 and staked a claim on Koranga himself. As a miner Levien was moderately successful, but his heart was elsewhere. From the start he believed that there was big money to be made on the Bulolo, but that loners like Sharkeye and Royal weren't going to make it. Levien reckoned—correctly as it turned out, though he was neither a miner nor an engineer or geologist—that the Bulolo gold had been washed down over centuries from Kaindi and neighboring peaks and that most of it had been carried farther downstream and deposited on the valley floor, the Bulolo flats. Unfortunately for the independent miners, that meant that instead of being concentrated in narrow streambeds, the bulk of Kaindi's gold was dispersed so thinly over so large an area that it could not be extracted manually. Levien wanted to form a company that could afford to exploit those deposits with dredging equipment.

In May 1926 Levien and some Australian financiers formed Guinea Gold No Liability, with the intention of dredging the Bulolo flats. From then on, the days of the lone prospector were numbered. One of them later wrote that by 1927, "the gold rush was over, and mining was entering on a new phase—that of big business. The talk was now mostly about dredging leases rather than claims. . . . Large, low-grade deposits of alluvial gold along the larger streams, that could be worked with modern machinery handling hundreds of tons of gravel a day, were of much greater importance than rich but limited deposits on the creeks."

Levien had another scheme up his sleeve. He wanted to fly in the supplies and equipment the new company would need. Hiring porters was expensive even for small-scale sluicing operations. To have native bearers carry goods into Bulolo from the coast cost as much as two shillings, fourpence per pound; air freight would cost threepence per pound. Besides, it would simply be physically impossible to carry the equipment needed to mine on the scale he envisioned. Levien got his way: A strip was cleared near Wau Creek, a tributary of the Bulolo, and in 1927 a DH37 biplane landed there, the first flight into the interior of New Guinea. The pilot, "Pard" Muster, later called Wau the "world's worst aerodrome." The landing strip was, and still is, on a steep slope surrounded by trees; the approach is between mountains that rise to ten thousand feet on one side and to eight thousand on the other and that are often enveloped in cloud to within a few hundred feet of the ground. Pilots must come in at the bottom of the slope, flying more or less straight at the hillside, giving the engine full throttle to get it uphill, and then swerving the plane sharply to one side to stop it.

Pard Muster's biplane could carry 600 pounds at a time. It was soon supplemented by two single-engine German Junker W-34 planes, capable of carrying 1,800 pounds apiece. Bit by bit the township of Wau emerged from pieces flown in from the coast: pigs, sheep, and cattle, billiard tables, pianolas, electricity generators, and a tennis court. The Wau Hotel arrived in sections. But there was no airplane that could carry the huge dredges needed to mine the flats. Even if they could be cut into sections—a considerable engineering challenge in itself—engineers calculated that the largest single piece would weigh more than three tons. But in 1929 the Junkers factory agreed to adapt two of their triple-engined G-31 planes to take loads of 7,000 pounds, 5,000 pounds more than the standard model could carry.

In 1930 the company, which had by then been rechristened Bulolo Gold Dredging (BGD) and bought by a Canadian company, Placer Development Ltd., cleared an aerodrome from the dense rainforest on the Bulolo so that the dredges could be flown directly to the flats. In one month in 1931 the Junkers brought 581 tons in over the mountains, more cargo than the combined airfleets of the world had carried in the previous twelve months. In 1932 the first dredge, weighing 1,100 tons, started work at Bulolo. Just two months earlier

Levien had died. His last words were "Placer shares will go to ten dollars."

The inaugural ceremony was a chance for self-congratulation. The administrator of the former German territory, a General Wisdom, was ecstatic: "It would seem as though some magician had waved a wand and given forth the fiat, 'Let there be gold' and there was gold. 'Let there be aeroplanes' and there were aeroplanes! 'Let there be a dredge' and there is a dredge. 'Let there be a happy community of workers' and there is one."

In 1933 this happy community included more than six thousand natives, indentured to their employers for three years. One quarter of them worked for BGD. An adult man received five to ten shillings a month, plus food and housing. Most had no idea of the value of cash and treated it simply as a novelty. Many spent it on airplane rides over the goldfields. Working conditions were difficult, especially in Edie Creek, where the cold was intense. The laborers could not get the foods to which they were accustomed, and disease spread rapidly through the work gangs. Dysentery was general and severe; one in eight of the natives who caught it died. Morobe Province, where the goldfields were, had by far the highest death rate in Papua New Guinea. When in 1933 it fell to 3 percent, it was still double the territorial average.

There were different views about the treatment of indentured laborers. An American called Ballock who visited the goldfields in the early 1930s said the laborers were kept in a kind of slavery and criticized the arrangements made for them. Workers in Edie Creek, for example, were issued only two blankets, when they needed at least four or five to keep adequately warm at night. Commissioner Dunstan complained in 1931 that native labor was being coddled and lengthened indentures from two years to three. One Australian writer pointed out, "Australians prided themselves on their standards of paternalism and . . . BGD aimed to create a contented community of indentured workers." BGD did not put their laborers in large barracks but housed them in groups of two, three, or four in huts. The men also cooked their own food and ate with their friends, instead of in a company canteen. Workers were encouraged to bring their families to the field, but very few did; the indenture system had severe effects on the families and villages the men left behind. After the war the indenture system, under which it was a criminal offense to break an employment agreement, was dropped. Working conditions at BGD deteriorated, however, with twelve men crowded into a hut measuring thirty by sixteen feet. Only one third of the men renewed their contracts. But, because there was a surplus of labor caused by the raising of the minimum urban wage to £3 a week, the company had no difficulty getting new men.

Eventually BGD had eight of the biggest dredges in the world working in the middle of the rainforest, or former rainforest. Far too heavy to move on solid ground, the dredges floated on man-made lakes. The largest dredge was several hundred feet long and over 50 feet high. They were in essence floating factories: Massive arms dug deep into the earth and deposited their loads into the treatment plant upstairs, where an assembly line of laborers toiled in shifts around the clock. As the gravel was processed and the gold extracted, the machine spewed out thousands of tons of sludgy waste and lumbered on. The huge scoops could go as far as 115 feet below ground to bring up great quantities of gravel, of which a small but precious proportion was gold. On the flats one hundred cubic yards of earth would yield little more than one ounce of gold, whereas miners on Edie or Koranga Creek sometimes got an ounce from one cubic yard. Nonetheless, with the dredges treating seven thousand cubic yards every day, the company brought in more gold than Sharkeye and his fellows had dreamed of. Between 1932 and 1942, when the war reached New Guinea, BGD recovered each year an average of 140,000 ounces of gold, worth $70 million at today's prices, and paid its shareholders regular annual dividends of $3 per $5 share.

Apart from 1942 to 1946, when the Japanese occupied New Guinea, the dredges worked night and day from 1932 until the mid-1960s. For twenty miles along the riverbed, they churned up the earth to a depth of one hundred feet or more until the flats were finally worked out. Afterward the company sold some of the dredges and simply abandoned the others. Following the river north from Wau to Bulolo, I came upon two of them. Rusting but upright, they dominated the narrow valley, incongruous there in the only habitat they had known. A family was living in one, sharing the shade of the giant factory with the few chickens they kept. They had come here, the father said, attracted by the gold.

. . .

With the dredges stopped, Bulolo is once again in the hands of the small miners. The difference now is that the New Guinea natives bending over the sluice boxes are working for themselves, not as indentured laborers to foreign employers. Morobe now has the highest proportion of migrants of all the mainland provinces, apart from the capital, Port Moresby. "Before independence the whites had this valley locked up; they kept the number of native squatters to a minimum. Now everyone and his brother is coming in for gold and the pace of deforestation has increased," one expatriate told me.

A few lucky miners make $400 a month, and some make much, much more. Wayne Gagne, a Canadian entomologist, who with his wife, Betsy, a botanist, spent several years at the Wau Ecology Institute, was in the bank at Wau one day when a barefoot miner came in and, speaking in pidgin English, told the clerk he wanted to withdraw all his money, $200, to pay for a party he was giving. The clerk consulted his records for a while before looking back up and saying nervously, "Well, you don't exactly have $200." "Okay," shrugged the miner, "just give me what I've got." The clerk disappeared into the back office. Very soon the manager emerged, smiling. "How much do you need?" he asked. "I want my money," the man repeated doggedly. "Would $200 be enough?" said the manager in an ingratiating tone. "You've actually got more than that, but we could give you $200 and then you'd have some money left in the bank. How would that be?" Evidently impatient with all the talk, the miner took the $200 and left. Afterward the clerk told Gagne that the man had $200,000 in his account and to have given it to him would have broken the bank.

But many of the newcomers have neither struck it rich on the goldfields nor been able to find good land for gardening. In 1975 Australia's Commonwealth Scientific and Industrial Research Organization conducted a soil survey of Papua New Guinea. The Wau and Bulolo valleys were classified as subject to severe erosion and not suitable for arable crops. With many people trying to farm on these steep slopes and fragile forest soils, and with many needless fires, kunai is spreading. Every new square foot of grassland means less land for gardening—and more hungry people. The women and children who sit listlessly by the side of the road are a new and unsettling element in the otherwise pleasant, and mainly expatriate, township of Wau, with its golf club and swimming pools. In those eyes one can catch a glimpse of the despair usually associated with urban hell-holes.

I didn't go to Bulolo looking for gold. I was drawn there by a reference in a logging industry journal to "a chopstick factory which produces over 40 million pairs of chopsticks per month for the Japanese market" from the forests of the Bulolo Valley. The company is owned by the multinational corporation Inchcape in partnership with the government of Papua New Guinea. When it was founded, in 1954, it was called Commonwealth New Guinea Timbers (CNGT) and was owned jointly by the Australian government and Bulolo Gold Dredging.

As the gold ran out, BGD looked for something else to do. The company had invested heavily in the Bulolo Valley. It had built an airport, several hydroelectric plants, roads, and sawmills. Bulolo, with its housing for natives and Europeans, its cinema, hospitals (one for natives, one for Europeans), shops, tennis courts, swimming pools, bowling green, rifle range, cricket pitch, billiard halls, and library, was a company town from top to bottom. So it made sense for BGD to stay in the area, even though shareholders accustomed to the high returns of gold mining would have to adapt to less spectacular annual dividends.

The company got permission to log Bulolo's magnificent araucaria forests. There were an estimated 1.6 million cubic yards of hoop and klinki pine in the valley, which were to be cut over a period of forty years. CNGT built a processing plant for turning the logs into plywood and veneers, mostly for export to Australia. The company was to log "selectively," leaving enough trees to allow the forest to recover naturally. The government, however, planned to establish plantations of hoop and klinki pines on certain of the logged-over areas, first clearing and burning the uncut trees.

By 1982 these plans had gone awry. Logging the natural forests was to have taken until 1994. At that point the oldest of the plantations would have been ready for the loggers to start on it. What actually happened was that the company ran out of natural forests ten or twelve years early. The reason, according to company officials, was that they had logged the araucaria forests faster than planned in order to keep up with high demand for wood. One forester I spoke to, however, called it "pretty poor forecasting" and suggested that the company had had to take more trees than anticipated because its logging techniques were wasteful.

The company's gross abuse of selective logging has crippled the forest's capacity for regeneration. "We were highly selective at first," Stan Barnes, PNG Forest Products' chief engineer and a longtime employee of the company, told me, "but standards gradually dropped; we went for smaller and smaller trees as it became feasible to produce thinner veneers." "That's true," Mike Callaghan, an Inchcape consultant assigned to the company, said, laughing. "Lots of these areas we have been through three or more times, in some cases five times. Each time getting a bit more desperate."

To make things worse, it was suddenly apparent that for several reasons the plantations would not be ready on time. In the early years a high proportion of klinki seedlings died. Eventually the problem was corrected by adding sulfur to the water used on seedbeds, but the problems disrupted the forestation program. More seriously, the plantations had not been pruned and thinned over the years. Thinning allows the remaining trees to grow thicker, not just taller; pruning gives a good straight trunk; together they produce trees with more usable timber. But deprived of that care, the Bulolo plantations did not have enough clear wood to make logging them worthwhile.

The company and the government still can't agree who should have maintained the thirty square miles of plantations once they were established. Each blames the other for failing to "follow up." Papua New Guinea's forestry law does not require logging companies to replant or encourage natural regeneration on land they have harvested. Nor does the government charge companies for the reforestation work it does. "We felt," said Oscar Mamalai, assistant director of the Office of Forests, "that foreign companies should not be given responsibility for reforesting because they have no long-term interest in Papua New Guinea."

Reforestation is made especially difficult in Papua New Guinea by the country's system of land tenure. Ninety-seven percent of the land is communally owned by clans and tribal groups. "There is no such thing as selling land in Papua New Guinea. It's never been a traditional concept," said Joe Loreakena, a New Guinean working for a Japanese logging company. All negotiations for logging and reforestation leases take place with the whole clan or tribe, which makes for long and trying meetings much dreaded by businessmen accustomed to air-conditioned conference rooms. Tribal groups have been willing to grant logging rights in return for royalties, but are less likely to allow reforestation of a logged-over area. The government is meeting with resistance in its attempts to make lucrative logging leases conditional upon the granting of reforestation rights. People are afraid that if they give the government or a private company the right to plant and maintain trees on their land over a period of thirty years or longer, they may never get the land back, a fear reinforced by the experiences of colonial days, when many tribes did unwittingly lose their lands in deals with foreign plantation companies.

The company, now called PNG Forest Products, adamantly maintains that it is not obliged to create or maintain the plantations it plans to use. "Reforestation is done by the government," Callaghan said. "Though naturally we take quite an interest in what they do because we expect to be operating here for a long time. We inform them of areas we consider suitable for afforestation." When asked what the company would do if the government made more stringent demands on it, Callaghan replied, "At the moment, the government would find it very hard to get tough with logging companies because we'd just go down the drain. The cost of operating here is so high that we're all living close to the bone. We'd have to pack up if they wanted any more out of us."

Since 1980, when Inchcape took over, its loggers have gone farther and farther afield to find timber to make up for the dwindling supplies of araucaria and the immature plantations. PNG Forest Products is now logging hardwood species—oaks, southern beeches, and dipterocarps—higher up in the mountains and on the coast around Lae. Very little is known about the growth cycle of hardwood species, so it is impossible to replant them after they are cut. Instead the company is hoping that the forests it is logging will regenerate naturally. But this is unlikely, given the area's high population density and shortage of suitable farmland. Generally settlers move to an area as soon as it is logged and burn the leftover vegetation so that they can farm there. Plantations are just as likely to be burned, so the government is experimenting with planting foreign pines that can survive fires and that grow well on fire-created grasslands. The aim is to use these species, whose timber is not particularly valuable, to rehabilitate the soil so that araucarias can be planted later, though the work is still at the experimental stage.

Burning logged forests has become so common in Bulolo that the government decided in 1982 to let the company extract trees three inches below the stated government guideline of seventeen inches diameter measured at breast height (DBH). "We've authorized them to cut smaller trees," Mamalai said, "because if the trees are left standing, they will only be burned anyway." One result is that if these forests do survive the slash-and-burn farmers, the surviving trees will take longer to grow back to a size at which logging would be economically feasible. No one knows yet whether the "selectively" logged forests will survive to be logged again. If not, the timber industry will turn again to virgin forests for its raw materials, leaving the logged forests behind—damaged, unproductive, and vulnerable to fire.

Besides having the only plywood factory, the only veneer plant, and the only chopstick factory in Papua New Guinea, PNG Forest Products is the country's biggest forestry employer, with 1,800 workers—and the second largest timber harvesting operation on the mainland. Local tribes get about $73,000 a year in royalties for timber the company takes from their lands. "Sometimes the government is slow with the royalties to the locals," Callaghan said. "When that happens, the people just stop the logging operations. It's happened a few times. They just put a barricade across the road and inform us they don't want us around. We just have to go along with it. Sometimes they close things for weeks at a time."

In 1981, the last year for which figures are available, the company harvested 250,000 cubic yards of wood. Over half of this was wasted in the process of turning it into 120,000 cubic yards of plywood, veneer, and other timber products. Also in 1981 Japan imported 480 million pairs of PNG Forest Products' disposable chopsticks, made from thinnings from the plantations. Wasn't it a bit frivolous, I asked Stan Barnes, to turn irreplaceable natural resources into disposable chopsticks? How about reusable chopsticks, as a first step to less wastage? "Well," he said, "the Japanese are supposed to be a very clean people. Now you might argue about that in some respects, but as far as this goes, they are certainly very fastidious. Most of these go to restaurants, and people want a fresh pair every time."

But PNG Forest Products will no longer be taking its share (about 5 percent) of the Japanese disposable chopstick market. The chopstick factory has been plagued by disaster since it started up in 1973. Soon after it opened, the company responded to a government-imposed tripling of the minimum wage by shutting down for some months. Then in September 1982, a fire ripped through the building, reducing it to rubble within an hour.

. . .

To look at Edie Creek now, it is hard to imagine that it was once a lush, unspoiled forest, home to such spectacular creatures as birds of paradise, cockatoos, birdwing butterflies, and the curious prickly anteaterlike creatures called echidnas. The miners have stripped whole mountainsides down to bare earth and deliberately eroded them in order to wash the gold they hope is there into the riverbed, where they can sluice for it. To satisfy demands for better access to gold country, the government is clearing crude roads to the few areas that are still inaccessible. The landscape is now a jumble of gravel heaps, prospectors' shacks, and diverted streams. There is a sad wash of color from the pale yellow of permanently bare hillsides, to the ochre of newly bulldozed roads, and the dirty brown of silt-filled rivers and streams. Kunai has pushed the forest back until trees grow only on the very crests. Close behind the bulldozers come the settlers. And they bring fire.

In November and early December of 1982 there were fires everywhere. Driving with the Gagnes on the Wau-Bulolo road past McAdam National Park, one of only two national parks in New Guinea, I spotted three fires. All were likely to spread, but since they were on the other side of a deep gorge, there seemed no way we could get to them. We spotted some children near two of the fires. They giggled when we warned, yelling across the chasm, that the blazes might get out of control. Invoking the name of the park ranger seemed to have some effect, and eventually some adults came along and said they would take charge, though they showed no concern about the fires.

The previous day the Gagnes had seen two small boys, sons of miners, set fire to the bush inside the park. Wayne made them put it out and then went to see the park ranger, Aitapo Ropase. He was still asleep, after having spent all night climbing in a remote part of the park to try and evict a family that was illegally gardening and setting fires there. The government doesn't provide a vehicle, so Ropase has to patrol the entire five-thousand-acre park on foot.

On the way back to Wau, the Gagnes assured me that many of the fires were started just for the hell of it. "If you ask people why they're burning, they can't give you a reason." I argued that there had to be a cultural explanation. Something that made sense to the burners, if not to me, or the Gagnes, or conservationists. I spoke to Harry Sakulas, the New Guinean in charge of the ecology institute. What was the real reason people started apparently senseless fires? "People burn to make gardens," said Harry. So there was sense to it? "Except when they just set fires for fun."

. . .

The popularity of burning in Papua New Guinea is not restricted to the Wau-Bulolo area. I took a bus north to Lae to talk to Jim Croft, a botanist at the national herbarium there. The four-hour, thirty-mile journey was a depressing panorama of coarsely grassed slopes. No gardens interrupted the smooth monotony of the deforested hills, but here and there plumes of smoke rose from behind a ridge, their sources hidden from view. "I don't think it happens in other parts of the world, but here people just love fires," said Croft. "People think that primitive people are always adapted to their environment, but I guess when you've got an infinite resource, you don't have to be adaptive. People are running out of land now, though. They're moving up the slopes throughout the country, but I don't think they've realized yet that it's turned sour. There is still a lot of rainforest in New Guinea. Thank God so much of it is so inaccessible." A week later, leaving New Guinea for Manila, I looked out the window as the airplane rose and counted twenty-five fires blazing below me before all the smoke blended together to make it impossible to distinguish one fire from its neighbor.

Notes

Bibliography

Index

Notes

14 "Goodbye": *Anthropology Resource Center (ARC) Bulletin,* Summer 1983, 7.
15 $13.5 million: Goodland, p. 62.
24 a second multimillion-dollar dam: Allen, pp. 318–42.
24 Ambuklao Dam: USAID, p. vi.
26 Nestor Yost: Quoted in Robin Wright, "The Great Carajás," *Anthropology Resource Center (ARC) Bulletin,* March 1983, 5.
27 "Indian nations within the GCP area": Ibid.
27 if it ever gets off the ground: Fearnside, "Land-Use Trends in the Brazilian Amazon Region," p. 144.
32 "Its lands are high": *Journal of Christopher Columbus,* trans. Cecil Jane (London: Anthony Blond, 1968), p. 192.
35 plants that grow in the Amazon: G. T. Prance, "The Origin and Evolution of the Amazon Flora," *Interciencia* 3, no. 4 (July–August 1978): 216.
35 *Symphonia:* Meggers, Ayensu, and Duckworth, p. 34.
35 tapirs: Ayensu, p. 73.
35 lungfish: Ibid., p. 80.
35 "Had I been able to obtain a passage": Wallace, *The Malay Archipelago,* vol. 1, p. 235.
35 "Eastern Hemisphere": Ibid.
35 "mark their limits": Whitmore, *Wallace's Line and Plate Tectonics,* p. 3.
35 "none of these occur": Wallace, *Malay Archipelago,* p. 22.
37 land links: Whitmore, 1981, p. 39.
37 the rainforest belt: Food and Agriculture Organization (FAO).
37 United Nations study from 1976: Lanly and Clement.
37 an even worse figure: Myers, *Conversion of Tropical Moist Forests,* p. 175.
38 most comprehensive study: FAO.
38 a few remnant patches: Myers, *Conversion of Tropical Moist Forests,* pp. 168–72.
38 "all accessible tropical forests": UNESCO, p. 55.

38 At present rates Brazil will lose: Mauro Reis, "The Problem of Combining Electricity and Forest Development," *Unasylva* 34, no. 137 (1982): 28.

39 iniquitous land distribution: World Bank, *World Development Report* (Washington, D.C., 1983), pp. 148–9; European Economic Community, *Basic Statistics of the Community* (Luxembourg, 1982), p. 17.

39 Taking potential farmland into account: Plumwood and Routley, p. 7.

39 4.5 percent . . . own 81 percent: Angarwal et al., p. 43; Eckholm, p. 10.

39 1 percent . . . own a third: Murdoch, p. 139.

40 by the top 8 percent: Murdoch, p. 139.

40 El Salvador: "One Member of El Salvador's Junta Who Became a Revolutionary," *The Times* (London), July 18, 1980.

40 Peru: Murdoch, p. 139.

40 "the Amazon is ours": *Newsweek,* October 20, 1980.

40 the chief culprit: Davis, pp. 148–9.

40 "low population density": Bethel, p. 217.

41 (FAO) figures: Quoted in Plumwood and Routley, p. 7.

41 exhausted by 1990: Myers, *Conversion of Tropical Moist Forests,* p. 169 and p. 157.

41 the Congo: *World Forest Action Bulletin* (Trelleck Road, Tintern, Gwent, U.K.), January 1984.

42 Pygmies: Turnbull, p. 21.

42 "victory upon victory": Mitchell, p. 73.

46 "where were the flowers?": Bates, p. 35.

47 tallest trees in the world: Richards, p. 3.

49 Some of these "tanks": Ayensu, p. 146.

50 more than five hundred feet: H. Synge, ed., *Biological Aspects of Rare Plant Conservation* (Chichester, England: John Wiley, 1981), p. 183.

50 Full sunlight hits the emergents: Longman and Jenik, p. 26.

52 three-toed sloth: J. K. Waage and R. Best, "Arthropod Associates of Sloths," in *The Evolution and Ecology of Sloths, Anteaters, and Armadillos,* ed. G. G. Montgomery (Washington, D.C.: Smithsonian Institution Press, 1982).

54 240 species: Ayensu, p. 122.

59 in terms of pure bulk: Jordan, p. 395.

59 species of trees per acre: Richards (1973), p. 60.

59 400 species: U.S. Interagency Task Force on Tropical Forests, p. 9.

59 Madagascar: UNESCO, p. 92.

59 New Caledonia: Myers, *Conversion of Tropical Moist Forests,* p. 89.

59 Quibdó: *National Geographic,* January 1983.

59 Panama: U.S. House Committee on Foreign Affairs, p. 74.

60 Malay Peninsula, Southeast Asia: Myers, *The Sinking Ark,* pp. 115–16.

60 500 resident bird species: Richards (1973), p. 60.

60 Mount Makiliang: Myers, *The Sinking Ark,* p. 23.

60 Ghillean Prance: Prance, pp. 195–213.

60 two thousand sweeps of a net: Richards (1973), p. 60.

60 a typical four-square-mile patch: National Academy of Sciences (NAS), *Ecological Aspects of Development in the Humid Tropics,* p. 55.

60 more than previously suspected: Richard Southwood, *Mankind and Ecosystems: Perturbation and Resilience* (Brisbane, Australia: Griffith University Press, 1983), p. 16.

60 Papua New Guinea . . . the Philippines: Myers, *The Sinking Ark,* p. 135.

61 birds . . . in Indonesia: Nigel Collar of the Conservation Monitoring Unit of the International Council for Bird Preservation (ICBP) in Cambridge kindly pointed this out to me.
61 birds of prey: International Council for Bird Preservation, *Proceedings of the World Conference on Birds of Prey* (Cambridge, England, 1982).
61 "the size of apple trees": Spruce, p. 256.
61 less than once in an acre: Hugh Iltis, *Extinction or Preservation* (Department of Botany, University of Wisconsin, 1978), p. 16.
61 A survey of five acres: R. F. Skillings and Nils O. Tcheyan, *Economic Development Prospects of the Amazon Region of Brazil* (School of Advanced International Studies, The Johns Hopkins University, 1979), p. 2.
61 A sixty-acre patch: Plumwood and Routley, p. 16.
62 "If the traveller notices": Wallace, *Tropical Nature,* p. 65.
62 still unknown to science: U.S. House Committee on Foreign Affairs, p. 75.
62 two thousand as yet unnamed species of fish: National Academy of Sciences (NAS), *Ecological Aspects of Development in the Humid Tropics,* p. 65.
63 the same process: Aubert de la Rue, p. 132.
63 many important crops: Sutlive, Altshuler, and Zamora, *Blowing in the Wind,* p. 315.
66 nutrients were in the biomass: Rafael Herrera et al., "How Human Activities Disturb the Nutrient Cycles of a Tropical Rainforest in Amazonia," *Ambio,* July 1982, p. 111.
67 litter layer: Richards (1981), p. 217.
67 newly laid lawn turf: Jordan, p. 396.
67 "the cornerstone of mineral conservation": NAS, *Ecological Aspects,* p. 134.
67 Puerto Rico: Farnsworth and Golley, p. 159.
68 absorbed by the root mat: Jordan, p. 397.
68 A tropical cloudburst: Marinelli, p. 5.
68 From half to three quarters of the rain: Bayard Webster, *IDRC Reports,* vol. 12, no. 4, 1984, pp. 17–18.
68 40 percent of farmers: World Wildlife Fund, *Tropical Rainforest Campaign,* p. A/4.
68 A single storm: Marinelli, p. 6.
69 washed down from the Himalayas: G. P. Hekstra, *The Green Revolution Confronted with the World Conservation Strategy* (Foundation for Ecological Development Alternatives, The Netherlands, 1981), p. 13.
69 In the Ivory Coast: UNESCO, p. 52.
69 "the average rate of sedimentation": Vohra, p. 3.
69 "a national emergency": Myers, *The Sinking Ark,* p. 259.
69 Twelve percent of the human race: U.S. Department of State, p. 17.
69 2.3 million people: *New York Times,* February 12, 1982.
69 annual floods: Vohra, p. 3.
69 Six billion tons of soil: Prescott-Allen and Prescott-Allen, p. 70.
69 twenty thousand were recorded in one day: Haigh, p. i.
70 raising riverbeds six to twelve inches: Prescott-Allen and Prescott-Allen, p. 71.
70 annual cost of damage: Vohra, p. 11.
70 $150-million pipeline: Haigh, p. 33.
70 CO_2: Council for Environmental Quality, *Global Energy Futures* (Washington, D.C.: U.S. Government Printing Office, 1981), p. 1.

70 as much as half of the added CO_2: George Woodwell, "The Carbon Dioxide Question," *Scientific American* 238 (1978).
71 rain in the Amazon Basin: T. Lovejoy and E. Salati, "Precipitating Change in Amazonia," in *The Dilemma of Amazonian Development,* ed. E. Moran (Westerly Press, 1982).
71 Panama: Mary Batten, "The Science of Life," *Science Digest,* July 1981, p. 118.
71 cloud forests: H. Vogelman, "Rain-making Forests," *Natural History* 85, no. 3 (1981): 22–5.
72 "most exuberant fertility": Bates, pp. 176–7.
72 "a rich and fertile land": Roosevelt, pp. 220, 279.
72 "green meadows and fertile plantations": Wallace, *Travels on the Amazon,* p. 334.
72 "There are mountains in Attica": Plato, *Critias,* iiiB.
74 Within forty or fifty years: Uhl, p. 75.
74 if the clearing is too large: UNESCO, p. 218.
74 they will not enter large clearings: Lovejoy and Schubart, p. 12.
74 One study counted: Uhl, p. 75.
74 well over one hundred years: Ibid., p. 78.
74 Angkor Wat: Myers, *The Sinking Ark,* pp. 118–19.
74 "exhausted the seed bank": Uhl, p. 78.
75 yields of maize: Richards (1981), p. 220.
75 Wallaba Forest in Guyana: Ibid.
76 "we know little or nothing": U.S. Interagency Task Force on Tropical Forests, p. 40.
76 a fair sample of rainforest life: Longman and Jenik, p. 7.
77 "The method I settled on": Donald Perry, "An Arboreal Naturalist Explores the Rain Forest's Mysterious Canopy," *Smithsonian,* June 1980, p. 46.
78 only two species: Richards (1981), p. 41.
79 "The tropical forest floras of Oceania are not very well known": UNESCO, p. 51.
79 a mere fourteen acres: Lovejoy and Schubart, p. 2.
79 100 that were entirely new to science: UNESCO, p. 94.
80 plant collections: Ghillean Prance, *South America* (New York: New York Botanical Garden, n.d.), p. 55.
80 unspoiled rainforest in Indonesia: Myers, *Conversion of Tropical Moist Forests,* p. 69.
80 Uttarakhand: Haigh, p. 6.
80 no response from Zaire: Myers, *Conversion of Tropical Moist Forests,* p. 9.
80 $40 million a year: U.S. House Committee on Foreign Affairs, p. 79.
81 scientific papers published each year: William J. Reimer, *Science in the Tropics* (Washington, D.C.: National Science Foundation, 1983), p. 3.
81 "no more than two dozen": U.S. House Committee on Foreign Affairs, p. 78.
82 tribal people are human beings: World Bank, *Tribal Peoples,* p. 36; Furneaux, p. 56.
83 they did not know they were acting wrongly: World Bank, *Tribal Peoples,* p. 37; *Survival International News,* no. 2 (1983), p. 8.
83 the Ayoreo Indians: Keefe; Holland.
84 Aborigines: Meggers, Ayensu, and Duckworth, pp. 340, 293.
85 human settlements in rainforests: Flenley, p. 119.
85 people . . . reached the lowland forests of the Amazon Basin: Meggers, Ayensu, and Duckworth, p. 313; Flenley, p. 119.

86 "These nations are so near each other": Markham, *Expeditions into the Valley of the Amazons,* p. 80.
86 completely destroyed within 150 years: Meggers, p. 121.
86 About half . . . are extinct: World Bank, p. 23.
86 Among the forest groups affected: Davis, p. 6.
86 Pygmies . . . Bantus: Yi-Fu Tuan, *Landscapes of Fear*.
88 an ability to store protein: Meggers, p. 27.
89 special nitrogen-fixing intestinal flora: UNESCO, p. 361.
89 Bekey . . . Silu: Ibid., p. 362.
89 "At first I was quite puzzled": Wallace, *Travels on the Amazon,* p. 171.
90 islanders were forced to wear clothes: World Bank, *Tribal Peoples,* p. 24.
90 the patrol salt lacked iodine: Ibid., p. 24.
90 "obvious signs of freedom": UNESCO, p. 380.
92 Andaman Islands: UNESCO, p. 440.
93 the main day-to-day authority: Lee Swepston, "The Indian in Latin America," *Buffalo Law Review* (Fall 1978): 723–4.
94 "beautiful ideals, but unrealistic": Quoted in Davis, p. 89.
98 "The Panare killed Jesus Christ": The story of the Panare is compellingly told by Norman Lewis in the (London) *Sunday Times Magazine,* May 15, 1983. The translation of the Bible story is taken from that article.
99 "Observe Auca daily activities": Peter Czura, "Ecuador," *Flying Colors* 11, no. 1: 46.
99 indigenous people are forbidden simple technologies: Swenson, p. 36.
99 "two prospector-geologists": ICL Research Team, *A Report on Tribal Minorities in Mindanao* (Manila: Regal Printing Company, 1979), quoted in Anti-Slavery Society, pp. 121–2.
99 "I am doing this for your own welfare": Anti-Slavery Society, p. 120.
100 Twenty-seven states: Convention 107, Concerning the Protection and Integration of Indigenous and Other Tribal and Semitribal Populations in Independent Countries.
102 "without their full and voluntary consent": World Bank, *Wildlife Management in World Bank Projects,* p. 65.
102 Tinggians . . . Cellophil: Anti-Slavery Society, pp. 74–120; Swenson, pp. 36–8.
102 "We are poor": Anti-Slavery Society, p. 61.
103 Txukarramaes: *Survival International Review* (Autumn–Winter 1980): 10, 27; Stuart Wavell, "Wounded Indians Fight for Their Land," *The Guardian,* October 15, 1980; Warren Hoge, "Brazil Indians Kill White 'Tresspassers,'" *International Herald Tribune,* October 7, 1980.
105 The Kuna: Catherine Caufield, "Can Scientific Tourists Save Panama's Rainforest?" *New Scientist,* June 24, 1982.
108 U.S. beef imports: Shane, p. 100.
108 one third of Costa Rica has been converted to pastureland: Parsons.
108 three and a half times as much beef: Shane, pp. 93, 89.
108 only twenty-eight pounds of beef a year: Myers, "The Hamburger Connection," p. 5.
109 two thirds . . . is now devoted to cattle: Nations and Komer, p. 16.
109 one quarter of all Central American forests: U.S. House Committee on Foreign Affairs, p. 80.
109 less beef now than they did: Shane, p. 89.

109 American's growing appetite: Myers, *Conversion of Tropical Moist Forests,* pp. 46–7.
109 Brazilian sausages and corned beef: Nations and Komer, p. 17.
110 Land ownership: Ibid., p. 14; Myers, "The Hamburger Connection," p. 7.
111 one animal per twelve acres: Fearnside, "Cattle Yield Production for the Trans-amazon Highway of Brazil," p. 223.
111 peasant agriculture can support up to one hundred: José Lutzenberger, address to Symposium on the Environment and the Future, May 3, 1982.
112 Weed control accounts for one quarter of the operating costs: Sutlive, Altshuler, and Zamora, *Where Have All the Flowers Gone?,* p. 94.
112 Nearly all . . . are now abandoned: Robert Goodland, *Environmental Ranking of Amazonian Development Projects* (World Bank, 1979), p. 29.
112 Among the companies doing so: G. T. Prance, "Exploitation and Conservation in Brazil," in *Systematic Botany, Plant Utilization and Biosphere Conservation,* ed. Inga Hedberg (Stockholm: Almqvist and Wiksell International, 1979), p. 147.
127 one of the only three: *Oryx* (Publication of the Fauna and Flora Preservation Society, London) 17, no. 4 (1983): 202.
128 at least two thousand years old: Meggers, Ayensu, and Duckworth, pp. 37, 158; Myers, *Conversion of Tropical Moist Forests,* p. 23.
128 In England: I am indebted to Richard Mabey for pointing this out to me. For a detailed discussion, see Christopher Taylor, *Fields in the English Landscape* (London: Dent, 1975).
129 Purely as a hunting and fishing ground: Furtado, p. 101; Klee, p. 107.
130 The Hanunoo people: *Hanunoo Agriculture: A Report on an Integral System of Shifting Cultivation in the Philippines,* Food and Agriculture Organization (FAO), Forestry Development Paper no. 12 (Rome, 1957).
131 "successful examples . . . are exceedingly rare": World Bank, *Tribal Peoples,* p. 14.
131 the Matses: Ibid.
132 forty thousand square miles of primary rainforest: Myers, *Conversion of Tropical Moist Forests,* pp. 24–5, 155.
132 Mudiyanse Tenakoon: *The Ecologist* 12, no. 5 (1982).
134 three to fourteen times more per acre: Angarwal, p. 18.
134 A 1979 study: Quoted in Murdoch, p. 151.
135 38 percent of Brazilians are malnourished: Jackson Diehl, "For Brazil, Agriculture May Be Miracle Waiting to Happen," *International Herald Tribune,* July 28, 1983.
135 Rubber is Malaysia's most important export: *The Times* (London, Malaysia supplement, August 30, 1982).
135 only 25 percent of the nation's farmlands: Myers, *The Sinking Ark,* p. 98.
135 The Lawa: Kunstadter, p. 12.
136 "incredible amounts of information concerning their environment": Grandstaff, p. 6.
138 At Yurimaguas: Sanchez et al.
138 Even though every dollar borrowed: Robert Goodland, Catherine Watson, and George Ledec, *Environmental Management in Tropical Agriculture* (Boulder, Colo.: Westview Press, 1984), p. 96.
140 called *chinampa:* Rex Keating, "The Chinampa," in *The Man and the Biosphere Program after 10 Years,* ed. UNESCO (Paris, 1982).
141 "food forests": D. B. Arkcoll, "Nutrient Recycling as an Alternative to Shifting

Cultivation," in *Proceedings of Conference on Ecodevelopment and Ecofarming, Berlin Science Foundation* (Oxford: Pergamon Press, 1979).

145 "large parts of Colombia": U.S. Department of Agriculture, p. 44.

145 twelve acres or more to feed on: Fearnside, p. 223.

147 silted up within fifteen years: Allen, pp. 318–42.

150 virtually all the hardwood logs: François Nectoux, Trade in Tropical Hardwoods (Earth Resources Research, London, unpublished), p. 17.

150 from Malaysia and Indonesia alone: Scott-Kemmis, p. 72.

150 keep almost all of it: Steinlin, p. 1.

151 exported more and more: Scott-Kemmis, p. 29.

151 Indonesia, for example: Bob Secter, "Deforestation Problem in Southeast Asia Seems Critical, Specialists Say," *International Herald Tribune*, March 26, 1982.

151 only about twice as much hardwood: Myers, *Conversion of Tropical Moist Forests*, p. 36.

151 Three quarters of their exports: Scott-Kemmis, p. 27.

151 the same low proportion: Ibid., p. 30.

151 $8.7 billion a year: John David Brazier, "Patterns, Trends, and Forecasts of Wood Consumption to the Year 2000" (Address to Annual Meeting of the British Association for the Advancement of Science, September 7, 1982), p. 3.

151 In Malaysia, Cameroon, and the Ivory Coast: Scott-Kemmis, p. 32.

151 50 percent of gross earnings: Ibid., p. 176.

152 transfer pricing: Ibid.

152 Nigeria was once a major timber exporter: *World Wood*, September 1983, p. 63.

152 The EEC, for example: Scott-Kemmis, p. 32.

152 U.S. subsidiaries in Latin America: Marinelli, p. 12.

153 a survey by the Brazilian government: Richard Bourne, *Assault on the Amazon* (London: Gollancz, 1978), p. 147.

153 A United Nations study: Scott-Kemmis, p. 175.

153 Brazil encourages companies to invest in the Amazon: Fearnside, "Land-Use Trends in the Brazilian Amazon Region," p. 142.

153 Unilever's logging operations there employ 450 people: Unilever, *Operations in the Solomon Islands* (London, 1980).

154 contract between Jant and the government of Papua New Guinea: Scott-Kemmis, Appendix I, p. 23; Oscar Mamalai, Office of Forests, Papua New Guinea, personal communication, November 1982.

154 "They should be doing reforestation": D. L. McNeil, *Negotiations with Transnational Companies in the Tropical Hardwood Sector* (Unilever, 1980), p. 34.

154 According to the United Nations: United Nations Economic and Social Commission for Asia and the Pacific, Feature no. F/01/80, 1980.

155 Japan itself has 60 percent of its land under forest: J. Westoby, *Halting Tropical Deforestation: The Role of Technology* (U.S. Congress, Office of Technology Assessment [in preparation]), Appendix II.

155 "Japan has a very clear strategy": Steinlin, September 7, 1982, personal communication.

158 "Thirty years ago we had thick mixed forest": Pandurang Ummayya and Bharat Dogra, "Planting Trees—Indian Villagers Take the Decision into Their Own Hands," *The Ecologist* 13, no. 5 (1983): 186.

159 "forestlands cover about 54 percent": Bureau of Forestry Development (BFD), p. 111.
159 in fact less than one third: Myers, *Conversion of Tropical Moist Forests,* pp. 95, 98; Percy Sajise, personal communication, November 24, 1982.
159 Every accessible forest: U.S. Agency for International Development (USAID), 1979, p. 4.
159 the number of timber licenses declined: BFD, p. 40; U.S. Department of Agriculture, p. 79.
160 logged out by the middle of this decade: Myers, *Conversion of Tropical Moist Forests,* p. 99.
160 "By the early 1990s": National Economic Development Authority, *Five Year Philippines Development Plan* (Manila, 1978), p. 89.
160 Most of the Philippines' remaining forests: BFD, p. 4.
160 biggest producer of wood and pulp in Asia: Fraser, "PICOP: New Losses Enforce Change," p. 21.
162 The most disruptive thing: Rao and Chandrasekharan, p. 12; Food and Agriculture Organization (FAO), p. 88.
163 A Filipino company that tried this system: USAID, 1980, p. 18.
163 for every ten that are deliberately felled: FAO, p. 88.
163 Only 33 percent were unharmed: U.S. Interagency Task Force, p. 18.
163 Research by one timber company: USAID, 1980, p. 18.
163 A study in Malaysia: Norman Myers, *Preservation and Production* (New York: Council on Economic Priorities, 1980).
164 4 cubic yards of wood per acre: FAO, pp. 87–8.
164 about twenty are exported to Europe: *Asian Timber,* March–April 1982, p. 29.
164 fewer than ten are accepted by foreign markets: Bethel et al., p. 243.
165 "small handful of tropical forest industrial companies": Fraser, "Testing Time for Asia's Forest Products Pioneer," p. 17.
166 twice as fast as it should be: USAID, 1980, p. 18.
166 roughly seventy-three cubic yards per acre: E. T. Tagudar, personal communication, November 24, 1982.
166 "a much greater amount than anticipated": Fraser, 1980, p. 14.
168 a survey in 1975: Scott-Kemmis, p. 100.
169 Queensland: *Rainforest Research in North Queensland* (Brisbane: Queensland Department of Forestry, 1983).
170 "after several decades of experimentation": Alfred Leslie, "Where Contradictory Theory and Practice Co-Exist," *Unasylva* 29, no. 115 (1977).
171 Many foresters are optimistic: John Spears, *Can the Wet Tropical Forest Survive?* (Washington, D.C.: World Bank, 1979), pp. 25–6.
171 Plantations also have a much greater yield: U.S. Department of State, p. 53.
171 now fewer than twenty million acres of industrial plantations: Duncan Poore, "World Forestry and Britain" (Lecture at Annual Meeting of British Association for the Advancement of Science, September 7, 1982), p. 8.
172 "In the Amazon": Eneas Salati and Peter Voge, "Depletion of Tropical Rain Forests," *Ambio* 12, no. 2 (1983).
173 "there are numerous examples of exotic species": Gary Hartshorn, *New Ecological Perspectives on the Management of Tropical Forests* (Hanover, N.H.: Institute of Current World Affairs, n.d.), p. 12.

173 the PICOP concession was uninhabited: A. F. Katibak, personal communication, November 24, 1982.
175 presidential decree (PD) 410: Anti-Slavery Society, pp. 153, 66–7.
177 "even forcibly removed": Ibid., p. 66.
178 "where slash and burn is discouraged": Fraser, 1981, p. 24.
179 it costs more than $2,500: USAID, "Background Research for Philippines Agroforestry Project" (Unpublished, n.d.), p. 5.
179 The interest costs of the loan: Deanna Donovan, *Forestry Developments in the Philippines* (Institute of Current World Affairs, 1978), p. 4.
180 a 1981 survey: E. L. Hyman, "Smallholder Tree Farming in the Philippines," *Unasylva* 35, no. 139 (1983).
184 "impressive system of irrigation": Wallace, *Malay Archipelago,* vol. 1, p. 236.
188 In parts of rural Java: World Bank, "Indonesia, Transmigration Program Review" (unpublished, 1981), vol. 1, p. 6.
188 One third of the land in Java: Murdoch, p. 139.
188 Plans for the program changed frequently: Perez-Sainz.
189 when Suharto took control of the government: A. A. Laquian, "Planned Population Redistribution: Lessons from Indonesia and Malaysia," *Habitat International* 6, no. 12 (1982): 41.
190 A 1976 study: Masri Singarimbun et al., *Transmigrants in South Kalimantan and South Sulawesi* (Yogyakarta, Indonesia: Population Insitute, Gadjah Mada University, 1977), p. 34.
192 a total of three medical clinics: Department of Information, *Transmigration in Development* (Republic of Indonesia, n.d.).
197 a Ministry of Agriculture study: David Butcher, personal communication, October 27, 1982.
198 migrants, or descendants of migrants: Soedjino, personal communication, October 27, 1982.
199 A report by the FAO: Food and Agriculture Organization (FAO), *Way Kambas Game Reserve Management Plan 1980/1–1984/5* (1979), p. 26.
199 nibong palms: Chris Humphries, British Museum, Natural History, personal communication, January 17, 1984.
209 malaria had struck many: Taylor, p. 75.
209 "A tree grows which they call the fever tree": Kreig, p. 171.
210 with no deaths from malaria: Ibid., p. 176.
211 "Opposition to Peruvian bark": Ibid., p. 178.
211 It revolutionized the art of medicine: Fredi Chiappelli, ed., *First Images of America* (Berkeley, Calif.: University of California Press, 1976).
211 "a very handsome tree": Spruce, vol. 2, p. 273.
211 "In the dry season it rained only in the afternoon": Kreig, p. 200.
214 "Though the trees are numerous": Taylor, p. 90.
214 "this highly esteemed product of the New World": Kreig, p. 190.
215 "Poor Manuel is dead also": Vietmayer, p. 101.
216 12.5 million pounds of dried bark they collected: Sutlive, Altshuler, and Zamora, *Blowing in the Wind,* p. 303.
216 there are serious problems with the synthetics: "A Deadly Disease Returns," *Newsweek,* October 25, 1982.
217 abandoned plantations . . . brought back into production: Vietmayer, p. 102.

219 Seventy percent of the three thousand plants: National Academy of Sciences (NAS), *Ecological Aspects of Development,* p. 89.
219 plants that contain alkaloids: Sutlive, Altshuler, and Zamora, *Blowing in the Wind,* p. 317.
219 who use it as an arrow poison: Ibid., p. 313.
219 "The arrow did not penetrate half a finger": Quoted in Kreig, p. 226.
219 "fortunate enough to find": Quoted in ibid., p. 229.
220 The antidote to curare is neostigmine: Sutlive, Altshuler, and Zamora, *Blowing in the Wind,* p. 288.
220 leukemia . . . Hodgkin's disease: Lewis and Elvin-Lewis, p. 131.
220 sixty alkaloids in this one plant: Humphreys, p. 516.
221 fifteen tons of periwinkle leaves: Cherry, p. 107.
221 Vincristine alone: Robin Levingston and Rogelio Zamora, "Medicine Trees of the Tropics," *Unasylva* 35, no. 140 (1983): 9.
221 "not even a chemist's wildest ravings": Humphreys, p. 516.
221 "Simply based on statistical probability": Cherry, p. 107.
221 A survey of 1,500 Costa Rican plants: U.S. Department of State, p. 22.
221 the Madagascar periwinkle was the only major anti-cancer drug: Mostapha Tolba, Address to United Nations Environment Programme's Session of a Special Character, Nairobi, 1982.
221 "the most versatile and available steroid raw material": Norman Applezweig. "Dioscorea, the Pill Crop," in *Crop Resources,* ed. D. S. Seigler (1977), pp. 149–63.
222 Tetradotoxin: E. Wade Davis, "The Ethnobiology of the Haitian Zombie," *Journal of Ethnopharmacology* 9, no. 1 (1983): 85–104.
222 Seven thousand medical compounds in the modern Western pharmacopeia: World Wildlife Fund, *Monthly Report,* no. 3 (1983): 7.
222 the value of such medicines: Prescott-Allen, p. 24.
223 natural medicines are the only sort most people will ever use: International Union for the Conservation of Nature, *Threatened Plants Committee Newsletter,* no. 9 (May 1982): 1.
223 The traditional Thai pharmacy: Wazeka, p. 28.
223 In the Malay Peninsula: World Wildlife Fund, *Tropical Rainforest Campaign, 1982,* p. A4.
223 more than 1,000 plants that Amazon Indian groups use: Cherry, p. 106; Mark Plotkin, *Conservation and Ethnobotany in Tropical South America* (Washington, D.C.: World Wildlife Fund-U.S., 1982), p. 12.
223 World Health Organization: Myers, *The Sinking Ark,* p. 127.
224 "can't take the aggravation": Quoted in Cherry, p. 108.
224 "the rather contemptuous attitude": Quoted in Richard Evans Schultes and Mark Plotkin, *Tropical Forests as Sources of New Biodynamic Compounds* (World Wildlife Fund, unpublished, 1982), p. 2.
225 *Cryptolepis sanguinolenta:* Prescott-Allen, p. 49.
226 The way epena is made: Richard Evans Schultes and Albert Hofmann, *Plants of the Gods* (London: Hutchinson, 1980), pp. 164–72.
227 Another ten crops: NAS, *Ecological Aspects,* pp. 78–83.
227 two hundred cultivars of their staple crop, taro: Morauta et al., p. 124.
227 every five, ten, or fifteen years: International Union for the Conservation of Nature, *World Conservation Strategy* (Gland, Switzerland, 1980), p. 3.

228 only two seeds . . . possessed the resistant gene: G. Khush, International Rice Research Institute, personal communication, December 10, 1982.
228 Cocoa plantations throughout the world: Prescott-Allen, pp. 59–62.
228 peanuts: NAS, *Ecological Aspects,* p. 83.
228 international trade in just nine rainforest spices: Wazeka, p. 28.
228 dependent on foreign sources of germ plasm: U.S. Department of State, p. 20.
229 a new species of citrus: Charlie Jarvis, "More Than Just a Pretty Face," *The Guardian,* September 11, 1980.
229 plants the academy says should be grown more widely: NAS, *Underexploited Tropical Plants.*
229 five hundred major insect pests are now resistant: R. P. Lamb, "The Daily Death of a Species," *Uniterra,* November–December 1981, p. 12.
229 "selects for its own failure": Robert Goodland, personal communication, October 1983.
230 importing three types of parasitic wasps: Myers, *The Sinking Ark,* p. 123.
230 the tachinid fly: Sutlive, Altshuler, and Zamora, *Blowing in the Wind,* p. 310.
230 kouprey: Brian Jackman, "Alive! The Ox from the Past," *Sunday Times* (London), August 22, 1982.
231 protein supply from wild animals: Gerardo Budowski, *The Socio-economic Effects of Forest Management* (Turrialba, Costa Rica: CATIE, 1981), p. 7; Prescott-Allen, pp. 21–3.
231 our understanding of tropical rivers: U.S. House Committee on Foreign Affairs, p. 76.
231 pescado blanco: IUCN, *World Conservation Strategy,* p. 3.
231 swamp forests of Cambodia's Tonle Sap: *IUCN Bulletin,* May 1980.
232 more than $20 million a year from forest product exports: Prescott-Allen, p. 42.
232 India's wild forest harvest: Norman Myers, "Forests for People," *New Scientist,* December 21, 1978.
232 a good stand of brazilnut trees: Sutlive, Altshuler, and Zamora, *Where Have All the Flowers Gone?,* p. 56.
232 use to stitch up their wounds: Ayensu, p. 123.
232 guar: NAS, *Underexploited Tropical Plants,* pp. 145–8.
232 *Thaumatococcus daniellii: Wild Genetic Resources* (London: Earthscan, 1982), p. 10.
233 the petroleum nut tree: Norman Myers, "The Four-Star Harvest," *The Guardian,* July 1, 1982.
233 ocelot coats: *The International Wildlife Trade* (London: Earthscan, 1979), p. 32.
233 five million cayman skins: Meggers, Ayensu, and Duckworth, p. 323.
234 Roosevelt's comments: Roosevelt, p. 220.
234 2 percent of the world's rainforests have been declared nature reserves: World Wildlife Fund, Switzerland, "Give Us Tropical Timbers," *Panda,* no. 3 (1980): 29.
234 Brazil has one of the most impressive: Anthony Rylands and Russell Mittermeier, "Parks, Reserves and Primate Conservation in Brazilian Amazon," *Oryx* 17, no. 2 (1982): 78–87.
235 "We have sixteen parks": Emil Salim (Address to Earthscan Seminar on Genetic Resources, Bali, October 1982).
235–6 "Why . . . is a pair of tigers more important?": I Made Tantra, I Made Sandy, and K. Kartawinata, "National Parks and Land Use Policy" (Address to World National Parks Conference, Bali, October 14, 1982).

236 Tai National Park: *World Wildlife Fund News,* March–April 1983; *IUCN Bulletin,* March–April 1981; Henri Dosso, Jean Louis Guillaumet, and Malcolm Hadley, "The Tai Project," *Ambio* 10, nos. 2–3 (1981).
237 Indonesia, where there are more than four hundred citizen nature groups: *Unasylva* 35, no. 140 (1983): 37.
237 "educational tourism" industry: Budowski, p. 4.
238 looking for gorillas: *World Wildlife Fund Monthly Report,* February 1982, pp. 53–7; Bill Weber, "People, Gorillas . . . or Both?" *Earthscan Feature* (London, July 1982).
240 "What difference does it make?": J. Wamsley, "The Vanishing Rainforest," *Pan Am Clipper,* June 1982.
240 "It is true that the Hepaticae": Quoted in Wallace's introduction to Spruce, vol. 1, p. xxxix.
244 75 percent of all the fish sold in Manaus: Goulding, *The Fishes and the Forest*, p. 253.
244 more than two hundred species in one lake alone: Agassiz quoted in Furneaux, p. 127.
244 of which only three thousand have so far been described: World Bank, *Wildlife Management,* p. 49.
244 "though poverty is at a distressing point": Goulding, p. 253.
245 the fish catch in Amazonian rivers fell by 25 percent: Michael Soulé (Address to Earthscan Seminar on Genetic Resources, Bali, October 1982).
245 The Uanano: Janet M. Chernela, "Indigenous Forest and Fish Management in the Uaupes Basin of Brazil," *Cultural Survival Quarterly* (Summer 1982): 17–18.
246 "the fisherman stations his canoe at dawn": Spruce, vol. 2, p. 381.
247 Dried pirarucu were the main rainy season protein food: Meggers, Ayensu, and Duckworth p. 323.
248 the busiest airport in the world: Sinclair, p. 126.
249 the German authorities began to allow . . . prospectors: Idriess, p. 16; Healy, p. 12.
249 William Dammkohler and Rudolf Oldorp: Healy, p. 13; Sinclair, p. 22.
250 "GOD SAVE 'UM KING": Idriess, p. 32.
250 "We crossed a bridge over a stream": Booth, p. 55.
251 "only two seasons": Booth, p. 123.
252 a total climb of twenty-nine thousand feet: Idriess, p. 134.
253 "Then I began to feel very queer": Booth, p. 73.
254 "Each plant, however insignificant, has its native name": Blackwood, p. 33.
255 "A platform is set up all around it": Ibid., p. 30.
256 "coarse, intractable grass": Ibid., p. 16.
256 "the gold rush was over": Leahy, p. 45.
257 air freight would cost threepence per pound: Healy, p. 74.
257 In one month in 1931: Idriess, p. 273.
258 "some magician had waved a wand": Wisdom quoted in Sinclair, p. 99.
258 "Australians prided themselves on their standards of paternalism": Healy, p. 83.
259 BGD recovered each year: Healy, p. 136.
260 Morobe now has the highest proportion of migrants: Office of Environment and Conservation, *The Effects of Population on Development* (Papua New Guinea, 1978).
264 In 1981, the last year: Office of Forests, Department of Primary Industry, *Facts and Figures 1982* (Konedobu, Papua New Guinea), p. 14.

Bibliography

Allen, Robert. "The Anchicaya Hydroelectric Project in Colombia." In *The Careless Technology,* edited by M. T. Farvar and J. P. Milton. London: Tom Stacey, 1973.

Angarwal, Anil, et al. *Life at the Margin.* London: Earthscan, 1979.

Anthropology Resource Center (37 Temple Place, Boston, Mass. 02111). "Hydroelectrics in Central and South America." *ARC Bulletin,* no. 11 (1982).

Anti-Slavery Society. *The Philippines.* 1983.

Arens, Richard, ed. *Genocide in Paraguay.* Philadelphia: Temple University Press, 1976.

Aspelin, Paul, and dos Santos, Silvio Coelho, eds. *Indian Areas Threatened by Hydroelectric Projects in Brazil.* Copenhagen: International Work Group for Indigenous Affairs, 1981.

Aubert de la Rue, Edgar. *The Tropics.* J. P. Harroy, 1957.

Ayensu, Edward, ed. *Jungles.* Jonathan Cape, London 1980.

Bates, Henry. *The Naturalist on the River Amazons.* London: Dent, 1969.

Bethel, James, et al. *The Role of the U.S. Multinational Corporations in Commercial Forestry Operations in the Tropics.* Seattle: University of Washington Press, 1982.

Blackwood, Beatrice. *The Kukukuku of the Upper Watut.* Oxford: Oxford University Press, 1978.

Booth, Doris. *Mountains, Gold, and Cannibals.* London: Cecil Palmer, 1929.

Bureau of Forestry Development. *1980 Philippines Forestry Statistics.* Manila.

Cherry, Laurence. "The Healing Power of Plants." *Bulletin of the Pacific Tropical Botanical Garden* 11, no. 4 (1981).

Davis, Shelton. *Victims of the Miracle.* Cambridge, England: Cambridge University Press, 1977.

Dressler, Robert. "Pollination by Euglossine Bees." *Evolution* 22 (1968): 202–10.

Eckholm, Erik. *The Dispossessed of the Earth.* Washington, D.C.: Worldwatch Institute (1776 Massachusetts Avenue, N.W.), 1979.

Farnsworth, E. G., and Golley, F. B., eds. *Fragile Ecosystems.* New York: Springer-Verlag, 1973.

Fearnside, Philip. "Cattle Yield Production for the Transamazon Highway of Brazil." *Interciencia* 4, no. 4 (1979).

———. "Land-Use Trends in the Brazilian Amazon Region." *Environmental Conservation* 10, no. 2 (1983).

Flenley, John. *The Equatorial Rain Forest: A Geological History.* Butterworth's, 1979.

Food and Agriculture Organization (FAO), Rome. *Tropical Forest Resources.* Forestry Paper no. 30 (1981).

Fraser, Hugh. "PICOP: New Losses Enforce Change." *World Wood,* January 1981.

———. "Testing Time for Asia's Forest Products Pioneer." *World Wood,* December 1980.

Furneaux, Robin. *The Amazon.* London: Hamish Hamilton, 1969.

Furtado, J. I., ed. *Tropical Ecology and Development.* International Society of Tropical Ecology, Kuala Lumpur, 1980.

Goodland, Robert. *Environmental Reconnaissance of Tucurui Hydro Project.* Eletronorte, Brasília, 1977.

Goulding, Michael. *The Fishes and the Forest.* Berkeley, Calif.: University of California Press, 1980.

Grandstaff, Terry. *Shifting Cultivation in Northern Thailand.* United Nations University, Japan, 1980.

Greenland, Dennis. "Bringing the Green Revolution to the Shifting Cultivator." *Science,* November 28, 1975.

Gressit, J. L., ed. *Biogeography and Ecology of New Guinea.* The Hague: Junk Publishers, 1982.

Haigh, M. J. *Deforestation and Disaster.* Address to British Association for the Advancement of Science, Annual Meeting 1983.

Healy, A. M. *Bulolo: A History of the Bulolo Region, New Guinea.* New Guinea Research Bulletin no. 15. Australia National University, Canberra, 1967.

Holland, Luke. *Indians, Missionaries and the Promised Land.* London: Survival International (29 Craven Street), 1980.

Humphreys, John. "Plants That Bring Health—or Death." *New Scientist,* February 25, 1982.

Idriess, Ion. *Gold Dust and Ashes.* Sydney, Australia: Angus and Robertson, 1935.

Janzen, Daniel. "Coevolution of Mutualism Between Ants and Acacias in Central America." *Evolution* 20, no. 3 (1966).

Jordan, Carl. "Amazon Rain Forests." *American Scientist,* July 1982.

Keefe, Norman. "Report on the Work Program with the Ayoreos." In *Economic Development at El Faro Moro: What Is Feasible?* Sanford, Florida: New Tribes Mission, 1977.

Klee, Gary. *World Systems of Traditional Resource Management.* London: V. H. Winston and Sons, 1980.

Kreig, Margaret. *Green Medicine.* London: Harrap, 1965.

Kunstadter, Peter. *Farmers in the Forest.* Honolulu: University of Hawaii Press, 1973.

Lanly, J. P., and Clement, J. *Present and Future Forest and Plantation Areas in the Tropics.* Rome: Food and Agriculture Organization (FAO), 1979.

Leahy, M., and Crain, M. *The Land That Time Forgot.* London: Hurst & Blackett, 1934.

Lewis, W. H., and Elvin-Lewis, M.P.F. *Medical Botany.* New York: John Wiley, 1977.

Longman, K. A., and Jenik, J. *Tropical Forest and Its Environment.* London: Longman, 1974.

Lovejoy, Thomas, and Schubart, Herbert. *The Ecology of Amazonian Development.* University of Cambridge, Center of Latin American Studies, 1979.

Lugo, A. E., and Brown, S. "Conversion of Tropical Moist Forests: A Critique." *Interciencia* 7, no. 2 (1982).

Marinelli, Janet. "Eco-Crime on the Equator." *Environmental Action,* March 1980.

Markham, Clements R., trans. and ed. *Expeditions into the Valley of the Amazons.* London: Hakluyt Society, 1859.

Meggers, Betty. *Amazonia.* Chicago: Aldine, 1971.

———; Ayensu, Edward; and Duckworth, Donald; eds. *Tropical Forest Ecosystems in Africa and South America.* Washington, D.C.: Smithsonian Institution Press, 1973.

Mitchell, John. "The Man Who Would Dam the Amazon." *Audubon Magazine,* March 1979.

Morauta, Louise; Pernetta, John; and Heaney, William; eds. *Traditional Conservation in Papua New Guinea.* Boroko, Papua New Guinea: Institute of Applied Social and Economic Research, 1980.

Murdoch, William. *The Poverty of Nations.* Baltimore: Johns Hopkins University Press, 1980.

Myers, Norman. *Conversion of Tropical Moist Forests.* Washington, D.C.: National Academy of Sciences, 1980.

———. "The Hamburger Connection." *Ambio* 10, no. 1 (1981).

———. *The Sinking Ark.* Oxford: Pergamon Press, 1979.

National Academy of Sciences (NAS). *Ecological Aspects of Development in the Humid Tropics.* Washington, D.C.: National Academy of Sciences, 1982.

———. *Underexploited Tropical Plants with Promising Economic Value.* Washington, D.C.: National Academy of Sciences, 1975.

Nations, James. "The Rainforest Farmers." *Pacific Discovery,* January–February 1981.

Nations, James, and Komer, Daniel. "Rainforests and the Hamburger Society." *Environment* 25, no. 3 (1983).

Nations, James, and Nigh, Ronald. "The Evolutionary Potential of Lacandon Maya Sustained-Yield Tropical Forest Agriculture." *Journal of Anthropological Research* 36, no. 1 (1980).

Parsons, James J. "Forest to Pasture: Development or Destruction?" *Revista de Biologia Tropical* 24 (supplement 1), 1976.

Perez-Sainz, J. P. *Transmigration and Accumulation in Indonesia.* Geneva: International Labor Office, 1979.

Plumwood, V., and Routley, R. "World Rainforest Destruction: The Social Factors." *The Ecologist* 12, no. 1 (January–February 1982).

Poore, Duncan. *Ecological Guidelines for Development in Tropical Rain Forests.* Gland, Switzerland: International Union for Conservation of Nature and Natural Resources (IUCN), 1976.

Prance, Ghillean, ed. *Extinction Is Forever.* Bronx, N.Y.: New York Botanical Garden, 1977.

Prescott-Allen, Robert, and Prescott-Allen, Christine. *What's Wildlife Worth?* London: Earthscan, 1982.

Rao, Y. S., and Chandrasekharan, C. "The State of Forestry in Asia and the Pacific." *Unasylva* 35, no. 140 (1983).

Raw Materials Group (P.O. Box 5195, Stockholm). *Raw Materials Report,* vol. 1, no. 2, 1982.

Richards, Paul. "The Tropical Rain Forest." *Scientific American* 229 (1973): 6.

———. *The Tropical Rain Forest.* Cambridge, England: Cambridge University Press, 1981.

Robinson, Michael. "Untangling Tropical Biology." *New Scientist,* May 3, 1979.

Roosevelt, Theodore. *Through the Brazilian Wilderness.* London: John Murray, 1914.

Sanchez, Pedro, et al. "Amazon Basin Soils: Management for Continuous Crop Production." *Science,* May 21, 1982.

Scott-Kemmis, Don. *Transnational Corporations and Tropical Industrial Forestry.* Science Policy Research Unit, University of Sussex, 1983.

Shane, Douglas. *Hoofprints on the Forest.* Washington, D.C.: U.S. Department of State, Office of Environmental Affairs, 1980.

Sinclair, James. *Wings of Gold.* Sydney, Australia: Pacific Publications, 1978.

Smith, John. *A Dictionary of Economic Plants.* New York: Macmillan, 1882 (N.B. not 1982!).

Spruce, Richard. *Notes of a Botanist on the Amazon and Andes.* New York: Macmillan, 1908.

Steinlin, H. J. "The Future Supply of Wood." Address to annual meeting of the British Association for the Advancement of Science, September 7, 1982.

Sutlive, Vinson; Altshuler, Nathan; and Zamora, Mario D.; eds. *Blowing in the Wind: Deforestation and Long-range Implications.* Williamsburg, Virginia: Studies in Third World Societies publication no. 14, College of William and Mary, 1981.

———. *Where Have All the Flowers Gone? Deforestation in the Third World.* Williamsburg, Virginia: Studies in Third World Societies publication no. 13, College of William and Mary, 1981.

Swenson, Sally, ed. *Background Documents Prepared for the Conference on Native Resource Control, Washington, D.C., October 12–15, 1982.* Boston: Anthropology Resource Center, 1982.

Swepston, Lee. "Latin American Approaches to the 'Indian Problem.'" *International Labour Review,* April–May 1978.

Taylor, Norman. *Plant Drugs That Changed the World.* London: Allen and Unwin, 1966.

Turnbull, Colin. *Wayward Servants.* London: Eyre and Spottiswoode, 1965.

Uhl, Christopher. "You Can Keep a Good Forest Down." *Natural History,* April 1983.

UNESCO, UNEP, FAO. *Tropical Forest Ecosystems.* Paris: UNESCO, 1978.

U.S. Agency for International Development (USAID). *Environment and Natural Resource Management in Developing Countries.* Washington, D.C.: U.S. Department of State, 1979.

———. "Report on USAID Mission to the Philippines." Unpublished, 1980.

U.S. Department of Agriculture. *Forestry Activities and Deforestation Problems in Developing Countries.* Washington, D.C.: USAID, U.S. Department of State, 1980.

U.S. Department of State. *Proceedings of the U.S. Strategy Conference on Tropical Deforestation, 1978.*

U.S. House Committee on Foreign Affairs. *Hearings on Tropical Deforestation,* 96th Cong. 2d sess., 1980.

U.S. Interagency Task Force on Tropical Forests. *The World's Tropical Forests: A Policy, Strategy, and Program for the United States.* Washington, D.C.: U.S. Department of State, 1980.

Vietmayer, Noel. "Incredible Odyssey of a Visionary Victorian Peddler." *Smithsonian,* August 1980.

Vohra, B. B. *A Policy for Land and Water.* Government of India, Department of Environment, 1980.

Wallace, A. R. *The Malay Archipelago.* London: Macmillan, 1869.

———. *Travels on the Amazon and Rio Negro.* London: Reeve and Co., 1853.

———. *Tropical Nature.* London: Macmillan, 1878.

Wazeka, Robert. "Bringing Modern Science into Herbal Medicine." *Unasylva* 34, no. 137 (1982).

Whitmore, T. C. *Tropical Rain Forests of the Far East.* Oxford: Clarendon Press, 1975.

———. *Wallace's Line and Plate Tectonics.* Oxford: Clarendon Press, 1981.

World Bank. *Tribal Peoples and Economic Development.* Washington, D.C.: World Bank, 1982.

———. *Wildlife Management in World Bank Projects.* Washington, D.C.: World Bank, 1983.

World Wildlife Fund. *Tropical Rainforest Campaign Document.* Gland, Switzerland: World Wildlife Fund, 1982

Yi-Fu Tuan. *Landscapes of Fear.* Oxford: Oxford University Press, 1980.

Index

Grateful acknowledgment is made to the following for permission to reprint previously published material:

The Anti-Slavery Society for the Protection of Human Rights: excerpts from *The Philippines: Authoritarian Government, Multinationals and Ancestral Lands*. Reprinted by permission of the Anti-Slavery Society for the Protection of Human Rights.

The Ecologist: excerpts from "Interview with Tenakoon," by Edward Goldsmith from *The Ecologist*, Volume 12, No. 5 (1982). Copyright *The Ecologist*, Camelford, Cornwall, UK. Reprinted by permisssion of *The Ecologist*.

Margaret Kreig: excerpts from *Green Medicine* by Margaret Kreig, 1965. Reprinted by permission of Margaret Kreig.

Macmillan Publishers Ltd.: excerpt from *Notes of a Botanist on the Amazon and Andes* by Richard Spruce. Reprinted by permission of the publisher.

Donald Perry: excerpt from "An Arboreal Naturalist Explores the Rain Forest's Mysterious Canopy" by Donald Perry. First published in the *Smithsonian* magazine, June 1980. Reprinted by permission of Donald Perry.

Times Newspapers Limited: extracts from article in The Sunday Times Magazine, by Norman Lewis, May 15, 1983. Reprinted by permission of the Times Newspapers Limited.